History of Technology

History of Technology

Fourth Annual Volume, 1979

Edited by

A. RUPERT HALL and NORMAN SMITH

Imperial College, London

MANSELL
London 1979

ISBN 0 7201 0916 7

ISSN 0307-5451

Mansell Publishing, 3 Bloomsbury Place, London WC1A 2QA

First published 1979

Articles appearing in this publication are abstracted and indexed in
Historical Abstracts and *America: History and Life*.

British Library Cataloguing in Publication Data

History of technology.
   4th annual volume: 1979
   1. Technology — History — Addresses, essays, lectures
   I. Hall, A. Rupert   II. Smith, Norman, *b. 1938*
   609   T15

ISBN 0-7201-0916-7
ISSN 0307-5451

Typeset by
Preface Ltd., Salisbury, Wiltshire
Printed in Great Britain by
The Scolar Press, Ilkley, West Yorkshire

Contents

Preface   v

P. S. BARDELL
Some Aspects of the History of Journal Bearings and Their Lubrication   1

K. R. FAIRCLOUGH
The Waltham Pound Lock   31

ROBERT FRIEDEL
Parkesine and Celluloid: The Failure and Success of the First Modern Plastic   45

J. G. JAMES
Iron Arched Bridge Designs in Pre-Revolutionary France   63

L. J. JONES
The Early History of Mechanical Harvesting   101

G. HOLLISTER-SHORT
The Sector and Chain: An Historical Enquiry   149

The Contributors   186

Preface

A good many years ago Dr Morris Kaufman, a graduate of the Department of History of Science and Technology at Imperial College, London, published a little history of the first plastic, celluloid; it is gratifying that a more recent graduate, Dr Robert Friedel, has returned to this theme. Another unusual subject is the history of lubrication, too long neglected, and of increasing importance in the nineteenth century as the size and speed of machines increased; the author of the paper, P. S. Bardell, is also a graduate of the Department. Yet a third innovation in the present volume is the printing of an early technical specification in the paper by K. R. Fairclough. Finally, it should be recorded that the paper by Dr Hollister-Short (another graduate of the Department) was drafted more than five years ago.

As before, we welcome contributions to this series and material to be considered should be addressed to the editors at the Department of History of Science and Technology, Sherfield Building, Imperial College, London SW7 2AZ.

A. RUPERT HALL
NORMAN A. F. SMITH

Some Aspects of the History of Journal Bearings and Their Lubrication

P.S. BARDELL

Introduction

Self-evidently the history of bearings can be said to date from the time when man first made use of simple rotational devices such as vehicle wheels, pulleys, potter's wheels, querns and so on. To what extent these early examples of bearings were lubricated is difficult to state but we can conjecture that a suitable substance for the purpose was very likely animal fat or even water, especially if mere heat dissipation was the primary objective. Despite the isolated appearance of novel ideas for friction-reducing devices which anticipated later developments,[1] for many centuries both the design of bearings and the practice of lubrication remained in a rudimentary state. As late as the sixteenth and seventeenth centuries bearings were generally little more than holes in the wooden or iron framework of a machine.[2] Although a plain hole in a machine frame or a block is the simplest type of bearing it is inadequate for most purposes as no provision is made for smooth running or for renewal of the bearing surfaces and by the beginning of the eighteenth century bearings with split bushes, commonly called 'brasses' or 'steps', retained by removable caps, were beginning to appear.[3]

It was the Industrial Revolution of the eighteenth century which led to the adoption of power-driven machinery on a large scale and provided the incentive for improvement. Prior to this the motive power for machinery was derived from the muscles of men or animals, or from water-wheels or windmills and interest in improving efficiency was slight. Speeds were relatively slow and the requirements of bearings were not severe. Occasionally in windmills the heat generated by friction caused a disastrous fire but as stone and wood were replaced by metal in bearings this hazard was reduced. Fats or water, depending upon the location and purpose continued to provide the lubrication and so effective was this arrangement that bearing elements often had a life of about a century.[4] The water-wheel played an important role in the early decades of the Industrial Revolution when it provided the only source of power for both the textile and the iron-working industries, and the second half of the eighteenth century and the first half of the nineteenth saw considerable achievements in improving their design and efficiency.[5] But, generally, bearings did not receive much attention and, apart from the progression to metal bearings, there was no reason why they should. There had not been any significant increase in the demands upon them and rotational speeds were still low, varying between 2.5 and 3.7 rev/min.[6] Within the mills themselves, however, there was a trend towards faster speeds and higher loads which focused attention on transmission methods and the importance of properly designed bearings together with the provision of adequate lubrication. There was moreover a growing awareness that inefficient transmission systems represented a loss of power and therefore of money. Indeed, the eminent engineer Sir William Fairbairn attributed his success in life 'to the saving of power effected by increasing threefold the velocity of the shafting in mills more than forty years ago'.[7]

Somewhat surprisingly, perhaps, the stationary steam engine did not directly stimulate the development of bearings. These low pressure, low speed, beam engines of simple and sturdy construction which were used almost throughout the whole of the nineteenth century were undemanding of their bearings although, once they were adapted to provide rotary motion, they were increasingly used to drive machinery. And, as with water-powered mills, it was the transmission systems which demanded, and received, investigation and improvement. At the beginning of the nineteenth century the drive to machinery was usually via ponderous, square section, cast iron shafts, often badly coupled and carrying large drums and pulleys. Sometimes the power required to keep them in motion was almost equal to that necessary to drive the machinery itself. Besides being costly to install and maintain such unsatisfactory arrangements also occupied considerable space and, through obstruction, deprived the factories of valuable light. Costs were reduced, power was saved and the efficiency of transmission systems was improved by the use of lighter section iron shafting accurately fitted and running at higher speeds. During the early decades of the nineteenth century there was considerable attention given to bearing design, testified by the patent literature of the period, especially for self-lubricating bearings and by mid-century the general form of the pedestal bearing or Plummer block had emerged and it has survived, substantially unaltered, to the present time. From the 1840s the brasses of these bearings were frequently lined with a white metal.

Unquestionably, the railways provided the greatest stimulus to the development of bearing design and operation during the major part of the nineteenth century and it is this subject with which this paper is largely concerned. The successful performance of bearings depended upon a satisfactory design, adequate supplies of lubricants with the necessary properties and the availability of suitable bearing metals.

The second half of the nineteenth century witnessed three main developments:

1. The achievement of a satisfactory axle-box design (and, later, the discovery of hydrodynamic lubrication).

2. The establishment of the mineral oil industry, initially to provide illuminating oils but fairly quickly developing into the major source of lubricants.

3. The production of good bearing alloys.

It will be convenient to deal with these topics separately.

The design and performance of railway axle bearings

EARLY FRICTION EXPERIMENTS

The relative motion of the two elements in a journal bearing is a sliding action and so friction experiments seemed relevant to achieving an improvement in bearing performance from the beginning of the railway era. Coulomb's experiments of 1781 were repeated and extended whilst G. Rennie concluded from his investigations, in 1829, that the friction of lubricated surfaces is determined by the nature of the lubricant rather than by that of the contacting surfaces.[8] Between 1831 and 1834, General A. Morin conducted a comprehensive series of experiments at the Conservatoire des Arts et Métiers on the friction between both unlubricated and lubricated surfaces.[9] Two points of interest emerge from the work of Morin. In the first place, his experiments were conducted on journals of from 5 cm to 10 cm diameter, a range typical of railway axle bearings and indicative therefore of his interest in this particular problem. Secondly, he devised his own recording dynamometer for measuring the frictional resistance in the bearings.[10] As the investigation of friction and lubrication continued throughout the century more and more special machines and instruments were introduced to assist in this work and some of them, efflux viscometers for example, are still used.

Although frequently quoted, the results obtained by these early workers were of comparatively little value because they were derived from tests covering only a limited range of conditions which were not commonly encountered in engineering practice. Only much later in the century was it appreciated to what a large extent friction was influenced by pressure, velocity and temperature. Nevertheless, the early work on the friction of journal bearings with a variety of lubricants, both continuously and intermittently applied, was significant because it was an approach to the subject which was to be repeated by many engineers and ultimately it led to some important conclusions during the 1880s.

Even earlier, in 1831, Nicholas Wood had determined the coefficient of friction on old, well worn, axles and found it to be about 0.02, a figure much lower than any quoted by Morin but considerably higher than values obtained in later German experiments.[11] On the other hand there was a good degree of correlation between the figures of Morin and those obtained by Mr Samuel Webber in measuring the frictional resistance of mill-shafting (where the loadings are not very high).[12] However, it is not really possible to draw any conclusions from their results because they were obtained under conditions inadequately specified. Furthermore this situation can hardly be regarded as remarkable for even today, when considerably more is known about the geometry and working of a hydrodynamic bearing, it is acknowledged that the difficulty of performing reproducible experiments and simultaneously measuring the important variables imposes limitations on the comparisons that can be made.

EARLY RAILWAY EXPERIENCE

Within a year or two of the opening of the Stockton and Darlington Railway the inadequacy of the axles and bearings of the wagons became obvious.[13] Wrought-iron axles, $2\frac{5}{8}$ inches diameter where they run in the bearings, had proved incapable of supporting the loads and had bent. Simple, open, half-bearings of cast iron were also unsatisfactory because they were not able to retain the lubricant even though trials had shown them to be more effective so far as frictional losses were concerned than brass bearings. Furthermore, it was also discovered that unless the bearings were at least 4 inches long they 'cut slight furrows' in the axle. These troubles led to new designs and further trials. One idea was to house the axles in cast iron tubes provided with a bearing at each end. Whilst a hole for the lubricant was cut in the middle of the tube no device to prevent its escape was provided at the ends. This arrangement added considerably to the weight of wagons without significantly improving their performance and after building only a few the design was discarded. The engineers realized that not only must lubricant be applied to the bearing surfaces it must also be retained within the bearing for reasons of economy as well as efficiency. Hence we see a progression to a bearing including a chamber to serve as an oil reservoir with leather discs fitted at either end of the bearing to form a seal. At this period the workshops of G. Stephenson in Newcastle were producing a similar design but with the added refinement of an iron ring on the axle and running in the reservoir to achieve a better distribution of the lubricant.[14] A clear parallel with self-lubricating pedestal bearings is discernible here, perhaps due to the influence of transmission system improvements in mills. Alternatively it might be that the early railway designs prompted the development of shaft bearing arrangements.

However, this design did not endure in railway work and during the 1830s typical bearings, rather than having the axle running through a reservoir, fed lubricant into the bearing via holes leading to the top of the shaft. Obviously the idea behind this arrangement was that gravity would draw the lubricant down and distribute it over the bearing surfaces. No doubt, at the time, it seemed sensible to introduce the lubricant to the loaded side of the journal for this is where it was most needed but many years later the work of Beauchamp Tower was to show this configuration was not the best. But even a quarter of a century before Tower's impressive and convincing demonstration, another engineer had already come to the same conclusion.

In 1853, Mr. W. Bridges Adams presented a paper 'On Railway Axle Lubrication' to the Institution of Mechanical Engineers. This paper was important for a number of reasons. It was the first one read to the Institute on this topic — in itself an acknowledgment of the importance of the subject. The problems of the railway in this respect he clearly defines; 'The well-made, case-hardened axles of a common road carriage are capable of running 5000 miles over a bad road with once oiling, with a small quantity of oil, while railway axles require greasing every 100 miles or less, with some few exceptions'.[15] Adams' survey and analysis of bearing design and lubrication is sensible and straightforward and based upon several years' experience in the field. The common practice of drilling the lubricating hole through the bearing brass made lubrication uncertain and the expedient of enlarging the size of these holes merely worsened the situation by reducing the bearing surface at the most important point, whilst the clogging of these holes stopped lubrication and caused overheating. Adams' solution to these problems was to apply the lubricant to as large a surface of the axle bearing as possible and this meant feeding it to the underside of the shaft. This could not be achieved with the commonly used open-bottomed bearings and so Adams designed a completely enclosed axle-box, which he patented in 1847, the bottom half of which was filled with lubricant so that the lower surface of the axle was bathed in it (Figure 1). As a further refinement the axle-box included two light wooden rollers which floated on top of the oil or grease and contacted the underside of the axle so that if the level of the lubricant fell below that of the axle lubrication would be maintained. Adams' design was a vast improvement on existing designs. By means of a flexible leather seal between the box and the axle he was able to prevent the ingress of dirt and grit. It was possible to make adjustments to compensate for wear of the bearing brass and a new brass could be fitted without lifting the carriage. When the paper was presented to the Institution its author was able to report that large numbers of the boxes had been made and had proved satisfactory in service. Adams also stated that although his initial design of 1847 appeared to be 'the original application of the principle', since then 'similar contrivances have been brought out by various other parties'. However, despite his sound idea of applying the lubricant to the underside of the journal Adams was not entirely convinced that feeding oil to the top was worthless and so he cautiously lubricated from both bottom and top.

By about 1850, then, a satisfactory plain bearing had emerged although the determination of its load-carrying capacity and frictional torque characteristics was not possible by any means other than experiment. Apparently, wherever the railway existed, feverish activity to investigate and offset frictional resistance was the order of the day. On the London, Brighton and South Coast Railway, work was carried out by Galton and Westinghouse with the cooperation of Mr W. Stroudley, the

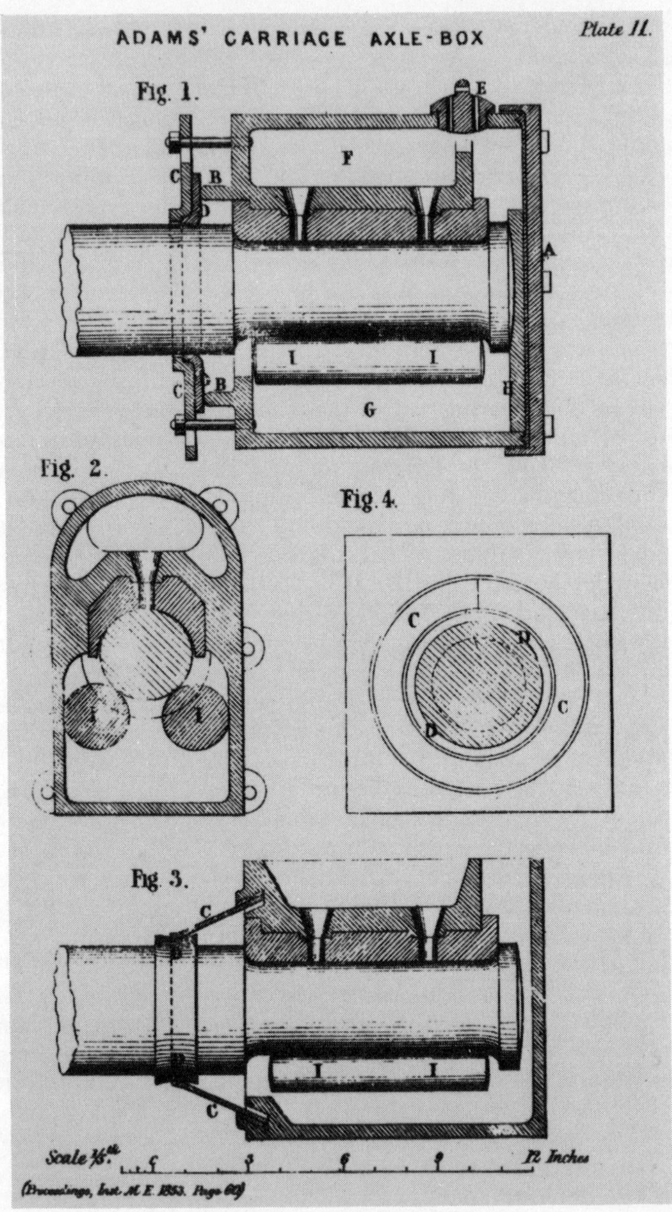

Figure 1. Adams' carriage axle-box (from 'On Railway Axle Lubrication' *Proc. I. Mech. E.* 1853 and reproduced here by permission of the Institution of Mechanical Engineers).

Company's locomotive superintendent. Stroudley himself was also engaged on experiments with the axles of rolling stock and locomotives under conditions similar to those encountered in practice.[16] In France, on the Paris and Lyons railway, tests were conducted by Poirée.[17] Similar investigations were being carried out in North America and in 1884 a paper was read before the American Society of Civil Engineers by Mr A.M. Wellington who reported on tests performed, in 1878, with loaded trucks.[18]

Unfortunately for the many engineers involved the results of the numerous experiments conducted led to greater confusion instead of clarification of the issues being examined. Thus the values of coefficient of friction given by Poirée were nearly twice as great as those given by Galton and Westinghouse. That friction diminished very rapidly as the speed increased was, however, a fact upon which both sets of results agreed but this observation was contradicted by the findings of others whose experiments showed that the friction increased with speed. Likewise, the effect of load appeared to be equally ambiguous.

Today, with our understanding of the curve (Figure 2) representing

$$\mu - \frac{\eta \omega}{\mathrm{p}}$$

and knowledge of the hydrodynamic, mixed-film, and boundary régimes of lubrication, it is relatively easy to see why the difficulties were encountered and to understand the perplexity caused by the many, evidently inconsistent, results.

Dissatisfaction with this confused state of affairs led Robert H. Thurston in the U.S.A. to undertake his own investigation of the subject. For, as he claimed, nothing of any real significance had been done since General Morin's experiments, especially to determine the amount of friction under the usual conditions then pertaining, that is, at working pressures of 500–1000 lbf/in^2. The figures quoted in the standard handbooks were of little use for as Thurston remarked 'No one can believe that the coefficients . . . given by the leading authorities . . . can be even approximately correct for these heavy loads'.[19] In his attempt to make good this deficiency he designed a machine that not only would 'exhibit the heating of a lubricated journal at pressures and speeds variable at will, but one that should also give with great accuracy, and at the same time, the more delicate but much more important measures of the amount of friction'.[20] The detailed design of this machine was carried out as a project by Mr J.A. Henderson, a student of Professor Thurston's at the Stevens Institute of Technology during 1873. It was the first such testing apparatus in which all the conditions of actual practice could be simulated and proved sufficiently successful for an enlarged version to be designed especially for railway work. This accommodated the standard railway wagon axle journal, $3\frac{1}{4}$ inches diameter and 7 inches long, and provided for

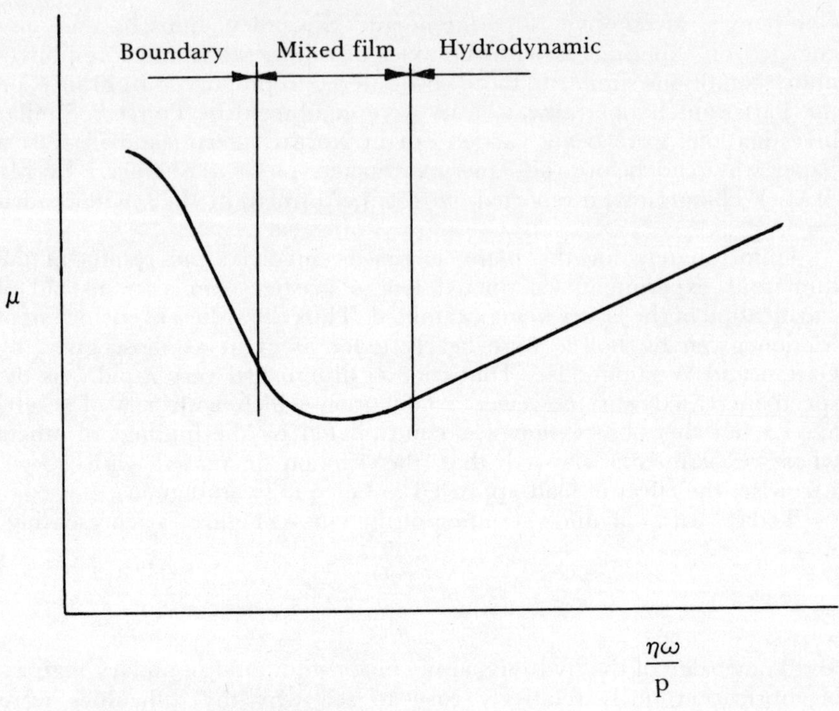

$$\frac{\eta\omega}{p}$$

Figure 2. Lubricating regimes.

μ = coefficient of friction ω = angular velocity of journal

η = viscosity of lubricant p = average bearing pressure

In the hydrodynamic regime an increase in friction will generate a higher temperature which, in turn, causes a viscosity decrease. This reduces $\eta\omega/p$ and causes a reduction in friction. Thus, this is stable lubrication because variations in the quantities are self-correcting.

A decrease in viscosity in the boundary regime would increase the friction. A temperature rise would follow and the viscosity be reduced still more. Thus the action is cumulative and leads to rapid heating and seizure.

The mixed film regime is a transition condition in which the bearing surfaces are separated from each other, partially by hydrodynamic forces and partially by thin layers of lubricant adhering to the surfaces.

a speed range varying between sixty miles per hour for a truck wheel of 26 inches diameter down to that of a 42 inch wheel running at 15 miles per hour. Similarly, the pressures were adjustable from a few pounds per square inch up to 400 lbf/in², or a total journal load of nearly 10,000 lbf. As soon as one of these machines had been built for the Stevens Institute of Technology others were manufactured, for the Pennsylvania Railroad laboratory at Altoona and for other railway companies, not only in the U.S.A. but in Europe too, where Messrs. W.A. Bailey & Co. Ltd, Albion Works, Salford, Manchester were the sole makers.[21] The sale of these machines in North America and across the continent of Europe is in itself evidence of the keen interest in journal bearing performance throughout the industrialized and industrializing countries of the world.

Thurston himself was responsible for conducting many hundreds of experiments but his results, published in 1878, although derived from a much more systematic and thorough approach to the problem than any earlier work, still exhibited inconsistencies. The implications of his findings were not understood and so it is hardly surprising that his results were accepted with reserve; indeed, they incorporated all the earlier apparent contradictions. At a constant bearing velocity (of 150 ft/min.) the coefficient of friction fell rapidly from 0.013 to 0.004 as the pressure increased from 50 lbf/in^2 to 500 lbf/in^2 and then rose again so that at 1000 lbf/in^2 it had about the same value as at 100 lbf/in^2.[22] Similarly, the frictional resistance was found to decrease with increasing velocity 'until the law changes'[23] at which point the resistance was found to increase with increasing velocity. A number of empirical formulae, relating the coefficient of friction, pressure, velocity and temperature, were given by Thurston but their use in machine design must have been of doubtful value. Of far greater assistance was the sound advice he gave on the practical aspects of lubrication and which, in summary, amounted to selecting the lubricant by actual tests upon a journal under pressures and velocities to be encountered in service. Economy was generally obtained by employing the lubricant which had been shown by experience to be best for the particular application. Unequivocally, he warned his readers that they could not afford to use inferior lubricants, even if they were freely available.

THE INVESTIGATIONS OF MR BEAUCHAMP TOWER

Whilst the more theoretical aspects of Thurston's findings make sense to us because we can see that his results showed (and showed for the first time) that the curve of friction against speed (Figure 2) possessed a minimum, they meant little to his contemporaries so far as throwing any light on the existing state of confusion was concerned. In these circumstances and in the year of publication of Thurston's results (in 1878), the Institution of Mechanical Engineers established a Committee on Friction which, in turn, commissioned Mr Beauchamp Tower to carry out an experimental investigation of journal bearing lubrication. This was carried out at the Edgware Road Works of the Metropolitan Railway. The first two reports of Tower, unquestionably his most important, were submitted to the Institution in 1883 and 1885.[24]

Attributing the considerable variation in bearing friction to the irregular action of the common methods of lubrication, Tower carried out his tests on a bearing immersed in an oil-bath which had the advantage of providing uniform and easily reproducible conditions (Figure 3). As his experiments

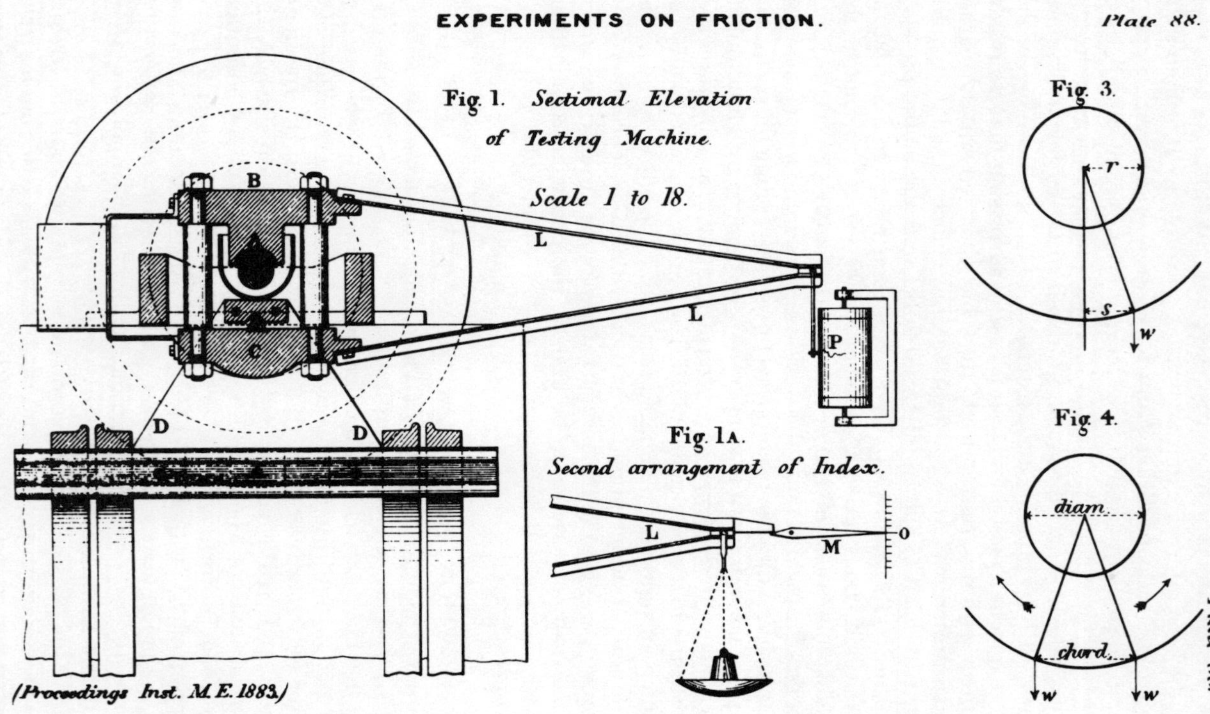

Figure 3. Experiment on friction (from 'First Report on Friction Experiments' *Proc. I. Mech. E.* 1883 and reproduced here by permission of the Institution of Mechanical Engineers).

showed, a full oil-bath was not a necessity and identical results were obtained when the bath was so empty that the bottom of the journal only just touched the oil. By means of gas jets under the oil-bath his apparatus also permitted an easy regulation of temperature.

Important findings came from these experiments:

1. It was clear that an abundant supply of oil was necessary to obtain repeatable measurements of friction torque.

2. The coefficient of friction far from being constant, as would be expected if the laws of dry friction applied, diminished as the bearing pressure increased and increased as the speed increased. From this Tower concluded, correctly, that in a properly lubricated journal the friction follows the laws of fluid friction more closely than those of solid friction.

3. There was a reduction of friction as the temperature increased.

4. Friction torque varied with different lubricants.

For the sake of comparison Tower tested his bearing with other systems of lubrication then in general use, including grooves of many patterns, but none of these proved to be so effective as the oil-bath. It was at this point that Tower made a discovery of profound importance. One of the bearings had been drilled to take a lubricator and when it was tested in conjunction with the oil-bath the oil welled up in the hole and overflowed. To prevent this inconvenience a cork was inserted in the hole only to be pushed out. The same happened with a more firmly fitted wooden plug. When a pressure gauge calibrated to 200 lbf/in² was screwed into the hole and the machine run the pointer went off the scale. Since the mean projected load on the journal was only 100 lbf/in² the experiment provided conclusive evidence that the brass was floating on a film of oil and also that the central pressure was more than double the mean pressure. Thus, Tower had shown that the journal was acting as its own pump and feeding oil round the bearing to the point of greatest pressure where the film of lubricant created carried the load (Figure 4). From this quite accidental discovery it followed that the correct place to oil a bearing was the point of minimum pressure, the motion of the shaft itself being sufficient to carry the lubricant to the region where it was most needed. Despite the enormous value of this discovery for the practical design of bearings, its relevance and significance were not recognized at the time and the discussion of the paper, which occupied two further meetings of the Institution, brought no comment from the participants on the topic.[25] There could no longer be any doubt, however, that the worst possible place to attempt to introduce oil to a bearing was at the centre of the seat of pressure but notwithstanding Tower's emphasis of this fact the practice continued for many years. One can only attribute this tardiness to the strength of tradition where, in situations not so demanding as the railways for example, the imperfect, intermittent and somewhat unreliable practice of supplying lubricant to the 'wrong place' was nevertheless adequate to ensure a fairly satisfactory mode of operation.

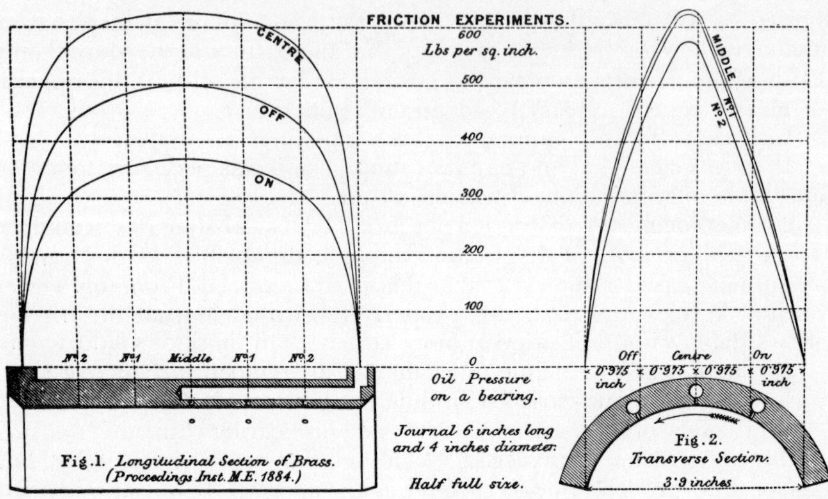

Figure 4. Pressure distribution curve obtained by Beauchamp Tower for a 155° partial bearing. After his discovery that the bearing brass was floating on an oil film, Tower conducted another set of experiments which showed that the pressure distribution around the bearing was of the form shown above. (From 'Second Report', *Proc. I. Mech. E.* January 1885 and reproduced here by permission of the Institution of Mechanical Engineers.)

The full implications of his findings were not recognized by Tower himself but the objectivity and accuracy with which he reported his results are characteristic of a great research engineer. Thus, he reported 'Early in the experiments it was found that, immediately after the motion of the shaft was reversed, the friction was greater than it was when the shaft had been running in the same direction some time'.[26] Following further detailed discussion of this point he offered his explanation. 'The phenomenon must be due to the surface fibres of the metal, which have been for some time stroked in one direction, meeting point to point and interlocking when the motion is reversed.'[27] Tower could not possibly have known the correct reason for this behaviour and his explanation, if not influenced by it, was certainly in the tradition of the mechanistic view of friction developed by de Belidor, Leonhard Euler and Coulomb. Nevertheless, his work, which departed completely from that of all his predecessors in so far as they confined themselves to the measurement of friction, showed that the load-carrying capacity of a bearing was a function of the relative surface velocity, the bearing surface area, the lubricant velocity and the oil film thickness. It unquestionably laid the foundations of a rational theory of lubrication. His achievement was not appreciated by all, however, for as a correspondent complained to an engineering journal in 1884, 'I had anticipated that Mr Tower's researches would have had some practical value; but as far as I can see they might just as well never have been undertaken . . . we gain absolutely nothing of any value from the discovery, if such it be, that the use of an oil bath diminishes friction enormously. We cannot use oil baths, and the fact is therefore of no importance.'[28]

HYDRODYNAMIC LUBRICATION EXPLAINED

Professor Osborne Reynolds concluded from Tower's experiments that the behaviour of the oil in the bearing should conform to hydrodynamic laws and, in 1886, published his analysis of Tower's results.[29] In his paper to the Royal Society, Reynolds showed that certain conditions were necessary for the generation of a film carrying a load by hydrodynamic action. These were:

1. A continuous supply of oil to the surfaces.

2. Relative motion of the bearing surfaces in a direction approximately tangential to the surfaces.

3. The ability of one of the surfaces to take up a small inclination to the other surface in the direction of the relative motion.

In showing the necessity of an oil film coverging in the direction of motion, Reynolds was indicating the positive role of a radial clearance. He was the first person to do this and in the case of Tower's journal bearing he attributed the clearance to a differential thermal expansion between the brass and the journal. Until this time, the function of a proper clearance between journal and bearing in contributing to the efficient performance of the assembly was, quite simply, not appreciated and any mention of dimensional control of the elements of the bearing is absent from the technical literature of the time. The need for accuracy and correct fit was, of course, realized and expressed. Thus W. J. M. Rankine stated that 'the accurate formation and fitting of bearing surfaces is of primary importance to the correct and efficient working of machines'[30] and R. H. Thurston told his readers that, having chosen the most efficient materials for the rubbing surfaces, 'they should be reduced to the most perfect state of smoothness and perfection in form and fit'.[31] However, the decade in which immense progress, in both theory and practice, was to be made was already at hand when Thurston's little book was published and when, nine years later the whole subject was given wider treatment in his *Friction and Lost Work in Machinery and Mill Work* (1887), the effect of clearance on frictional resistance was acknowledged and reference made to investigations of this phenomenon by Lasche.[32]

The variables designated by Tower as contributing to the load-carrying capacity of the bearing were formulated into mathematical terms by Reynolds but in order to do this (i.e. obtain the 'Reynolds' equation) a number of assumptions had to be made including the following:

1. Oil flow within the bearing clearance area is laminar.

2. The viscosity of the lubricant is constant.

3. Fluid inertia forces can be neglected.

Eccentricity, e = distance between shaft and bearing centres, B and A respectively, under operating conditions. Radial clearance in the bearing = R − r

Eccentricity ratio

$$= \frac{e}{R - r}$$

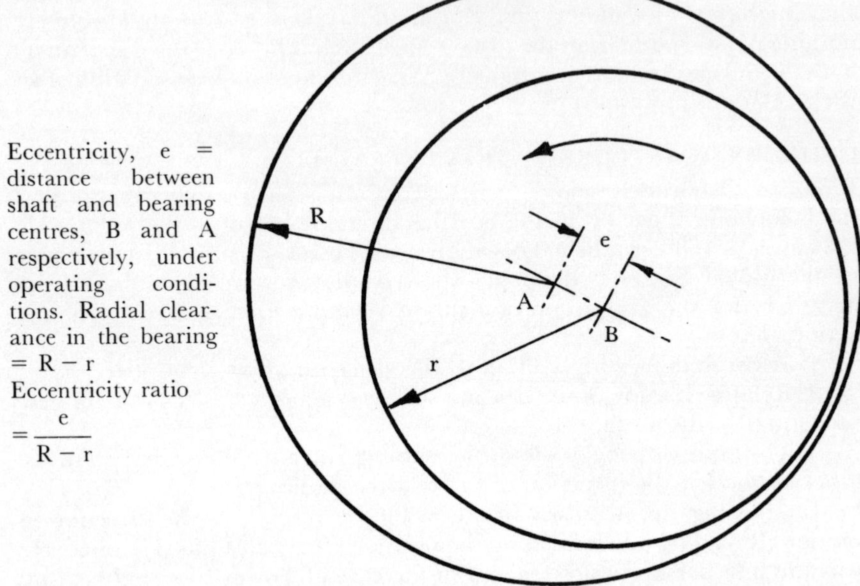

Figure 5. Eccentricity ratio. The eccentricity ratio defines the position taken up by the shaft within the clearance space and is an important factor in journal bearing design and operation. It is related to the geometry of the bearing as shown.

4. The oil film is continuous.

5. Variations of pressure with depth in the oil film can be completely neglected.

Reynolds' three-dimensional equation was not amenable to general solution and the close correspondence he achieved with Tower's experimental results was fortunate and due to his well-chosen assumptions. Because the values of eccentricity ratio (Figure 5) and radial clearance in Tower's experiment were unknown, the accuracy of Reynolds' analysis was dependent upon his assumptions and the quantitative limitations of his work is evident. As Professor A. Cameron has written, 'since this paper, at no time have theory and experiment agreed so well'.[33] Tower's tests were carried out on a bearing with a length/diameter ratio of $1\frac{1}{2}$, and an angular width of 157°. Reynolds however adapted the 'infinitely long' theory to them and in doing this he was simplifying the problem by reducing it to a two-dimensional one in which end leakage could be neglected. Modern industrial design practice tends to use length/diameter ratios of 1, or even less, which emphasizes the practical consequences of side flow. Even so Reynolds' simplified approach to the assessment of

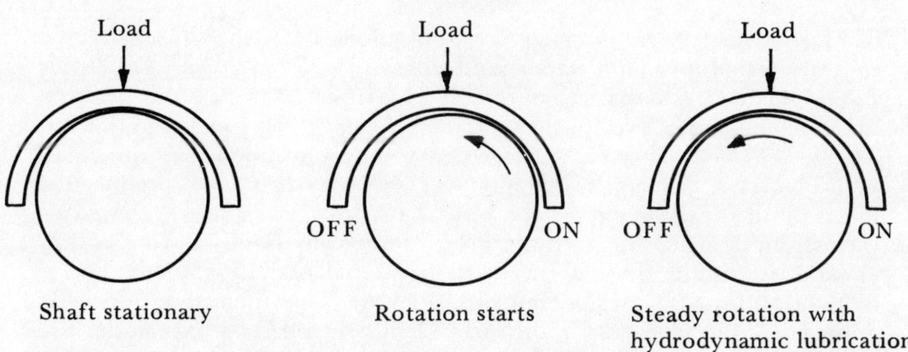

Figure 6. The generation of the hydrodynamic film. When the shaft is stationary the axes of both bearing and shaft will be in line; but when rotation begins the journal will ride up the bearing surface, on the 'ON' side, and in doing so will drag in some lubricant until it ultimately slips away and takes up a steady running position with the axis of the journal to the 'OFF' side of the line of load. The bearing is then entirely supported by the wedge of lubricant.

bearing performance has continued to be used and an allowance made for its tendency to give rather high bearing load capacities in terms of other parameters such as surface area, film thickness, etc.

The mathematical analysis of the journal bearing has, from the beginning, been subject to considerable limitations. Reynolds himself obtained solutions to his equation only for eccentricity ratios up to 0.5 and it was not until 1904 that A. Sommerfeld achieved a solution of the equation for a full 360° bearing and for all values of eccentricity ratio. But he, too, had to reduce the problem to a two-dimensional one by means of the concept of 'infinite length'. He also assumed that the oil film exerts a positive pressure around the full 360° of the bearing which is not confirmed experimentally. Without such assumptions a complete analytical solution, which is extremely complex, has not yet been achieved and even the approximate solutions, obtained for particular cases, do not lend themselves readily to industrial design. In a rather unexpected way Reynolds' theory contributed significantly to an appreciation of the action of a journal bearing and solved Tower's difficulty over the increase in friction upon reversal of motion of the shaft by showing that the point of nearest approach of the journal to the brass is offset from the line of load and, what is more, offset to the 'off' side of the line of load (Figure 6). Notwithstanding the quantitative limitations, referred to above, Reynolds' paper clearly demonstrated the qualitative validity of the hydrodynamic theory and it became fundamental to all subsequent work in the field.

At the other end of Europe, in St. Petersburg, simultaneously and quite independently, the principle of hydrodynamic lubrication was established by Nikolai Pavlovich Petroff, a Professor of Mechanics, who was involved with the engineering problems of Russia's rapidly expanding railways. His first major paper published in the *St. Petersburg Engineering Journal*, in 1883, was based not on any experimental work of his own but upon an analysis of previously published work. Petroff concluded that the journal and bearing were separated by a fluid film under certain conditions.[34] Using Newton's equation for viscous shear, he produced a mathematical expression for the bearing friction. His analysis, based on the assumption that the journal runs concentrically in the bearing, is only an approximation since if this was the actual operating mode hydrodynamic lubrication would not be achieved. Nevertheless, it is a good approximation, especially in the case of lightly loaded bearings, and is still used. His later work involved tests on animal, vegetable and mineral oils and appears, generally, to follow the pattern of exhaustive testings which characterize the period.

The 1880s were, then, very important in the history of plain bearings because during this decade not only was hydrodynamic lubrication discovered but its theoretical basis was also established.

SUBSEQUENT WORK

However useful mathematical analysis was in understanding the theoretical performance of bearings, its value to practical designers was limited. The incorporation of the results of scientific investigation into the design process depended upon an understanding of these results by those who should be using them in the course of their work. The need for making this understanding easier has been appreciated and voiced for many years. In the second discussion meeting on Tower's paper, in January 1884, Mr J.C. Fell, one of the members of the Institution's Committee on Friction, called for some kind of diagrammatic representation not only of Tower's results, but those of other experimental researchers as well, so that they would convey 'a clearer and more lasting impression'.[35] Over sixty years later, and in the very same Institution, the same plea was still echoing. In 1949, Mr D.B. Welbourn, commenting upon 'The full journal bearing' by A. Cameron and Mrs W.L. Wood, was still urging the need — if the paper were to be of use to 'ordinary designers' — for an appendix, to show in a non-mathematical form, the application of the curves and tables to the design of journal bearings.[36] And since the authors themselves had had to depend upon the Mathematics Division of the National Physics Laboratory to carry out their calculations it seems that Mr Welbourn, on behalf of ordinary designers, was not making an unreasonable request.

At the moment, two such design procedures have emerged. The older, and more abstruse, dating from the early 1930s, is due to A. Kingsbury[37] and S.J. Needs[38] and is based on the two-dimensional form of Sommerfeld's equations but modified by the introduction of appropriate leakage factors for a given length-to-diameter ratio. Perhaps mindful of the repeated pleas for a design method appropriate for use in the general industrial design office F.W. Ocvirk and G.B. Dubois produced, in the 1950s, a method utilizing charts, based on both analytical and experimental work, and a sequence of calculations whereby, for a chosen operating temperature, bearing performance data such as eccentricity ratio, oil film thickness, peak pressure of the oil film, friction torque, and oil flow can be determined.[39] The short interval between the discovery of hydrodynamic lubrication and its scientific explanation, followed by the considerably longer period before the results were presented in a form suitable for practical use suggests that adherence to traditional design procedures was not necessarily due to a reluctance to innovate on the part of engineers. Rather it was the result of a lack of awareness of any relevant theory. Inadequate communication, bemoaned by Messrs. J.C. Fell and D.B. Welbourn many years ago, is still, unfortunately, not uncommon.

ROLLING BEARINGS

A reduction in bearing friction could obviously be obtained if the sliding action of plain bearings was replaced by a rolling action. Concern with the provision of satisfactory plain bearings for rolling stock did not, therefore, prevent an interest being taken in ball and roller bearings. In fact it stimulated it as did the knowledge that these devices had been used successfully on road carriages since the late years of the eighteenth century. Between 1858 and 1873 three British patents were issued for axle-boxes employing either balls or rollers.[40] But however good these designs were, experiments with them were foredoomed to failure in this particular application because, due to the nature and magnitude of the stresses involved, only the best quality steels properly hardened and manufactured to a high degree of accuracy could be successfully used. Neither suitable steels nor the necessary production methods were available on a large enough scale until late in the century. It was in the bicycle that ball bearings first came into extensive use. With the introduction of the motor car, about 1895, whose production was initially in the hands of the cycle manufacturers,[41] attempts were made, by merely scaling up the elements, to adapt the 'cup and cone' bearing of the bicycle to the automobile. However, this type of bearing, although very successful under the relatively light loads and moderate speeds encountered in the bicycle proved entirely inadequate to cope with the heavier loads of cars and so many designers reverted to plain bearings.

At the beginning of the present century more experiments with ball-bearing axle-boxes were carried out but not with any great success because bearings of sufficient load-carrying capacity and accuracy of fit still could not be produced. Success was not achieved until shortly after the First World War during which considerable problems with road-wheel bearings had been encountered. From the beginning, continuous difficulty had been experienced with plain bearings and this led to development work, mainly in America, on rolling bearings. An outcome of this intensive effort was the tapered roller bearing, patented by

Henry Timken and R. Heinzelman in 1898, a device able to carry, simultaneously, both axial and radial loads and which, in addition, reduced starting resistance by as much as 60 to 70%.

Strangely enough, it was the application of the internal combustion engine to railway traction that led to further attempts to replace the plain journal bearing by roller bearings in axle-boxes. A point was reached where the starting capacity of the engine and transmission limited the size of carriage and this apparent impasse led to the realization that the use of roller bearings in axle-boxes would lead to a very marked improvement. To meet the requirements of the railway, a roller bearing axle-box had to sustain heavy loads at high speeds, severe hammer blows on the bearing surfaces and considerable axle end thrust, besides possessing such practical features as simplicity of design, efficient lubrication, and safe, reliable and economic operation with easy maintenance. As a result of a large amount of intensive development work both on the track and in the laboratory a successful bearing design was evolved and the manufacture of Timken Roller Bearing Axle-Boxes began in England during 1928.[42] Their success on rolling stock has led to their widespread adoption for this purpose throughout the world and brought about the disappearance of the plain journal bearing from the duty it performed so well and for so long.

Lubricants

THE GROWTH OF THE MINERAL OIL INDUSTRY

For the first three-quarters of the nineteenth century animal and vegetable oils were used almost exclusively for lubrication but by the end of the century they had largely been superseded by mineral oils. A fairly comprehensive and detailed survey of the lubricants used during the century has recently been made[43] and therefore this section will be confined to a brief review of the colossal demands of the rapidly expanding railway network, on a world-wide scale, for lubricating oils and the effect of this upon the development of the mineral oil industry.

THE BIRTH OF THE PETROLEUM INDUSTRY

The demand for improved illumination increased during the late eighteenth century, as a concomitant of the Industrial Revolution, and this demand was partially satisfied by the introduction and spread of coal-gas lighting. It could only be used however where connection to a gas main was possible and so was not available to the bulk of the population either in this country or in Europe. The most promising alternative to gas was the oil lamp which became a much improved source of light with Pierre Argand's introduction of the annular wick in 1782. Improvements in the oil lamp resulted in a demand for oil of better quality than that derived from animal and vegetable sources, and so began the search for a cheap, plentiful and better supply of illuminating oil. Petroleum offered the best prospect.

The modern petroleum industry emerged in the 1860s following the discovery of oil in Pennsylvania by Edwin L. Drake on 27 August 1859. It was the developing American industry which, for most of the following twenty-five years, produced the only significant supply of exportable oil. And, indeed, throughout this period the bulk of the output of the American refineries was exported — almost entirely to Europe.[44] Initially, the residue of 'heavy oil' resulting from the distillation of crude oil to obtain the illuminating oils was disposed of by burning as a waste product or, on an insignificant scale, it was distilled again to make lubricants. The industrial expansion of Western Europe and North America could not be sustained on animal and vegetable oils alone and this provided a stimulus for developing the production of lubricating oils. Transport was undoubtedly the most significant factor in the expanding industrial economies of the world and within this sector the growth of railways was the most prodigious. From a modest beginning in England, about 1830, railway systems were constructed and in operation all around the world, including China and Japan, by the mid-1880s. The following table[45] gives an indication of the growth of the railway in several European countries during the second half of the nineteenth century:

| | | | Year: | |
			1850	1900
Country:	Austria–Hungary	Miles of track:	980	22,580
	France		1,810	23,690
	Germany		3,640	32,120
	Great Britain		6,090	18,690
	Russia		310	33,090

In America too, this same rapid expansion took place and the fifteen years from 1884 saw an increase in the track network of about 62,000 miles.[46]

The global extension of the railways, the increase in the number and size of locomotives and rolling stock together with higher operating speeds created the need for considerably greater quantities of lubricating oils along with improvements in the quality. A measure of this phenomenal growth in demand is given by the export of lubricating oil from the United States, which rose from 1,244,300 gallons in 1874 to 67,424,400 gallons in 1899. Of these totals Europe took 1,166,100 gallons in 1874 and 53,722,300 gallons in 1899.[47] By the end of the century supplies to Western Europe were being augmented by imports from Russia; with the opening of the Baku oilfield during the 1880s the Russian oil industry emerged as a major producer and, indeed, in 1898 surpassed the United States in its output of crude oil. As with the American industry there was a distinct tendency for the proportion of lubricating oils produced to increase with time and, in fact, it rose from less than 16% to over 24% between the years 1884 and 1899: during this period exports to Western Europe rose from about 81,000 barrels, at the beginning, to almost 2,000,000 barrels at the end.[48]

THE PROPERTIES OF LUBRICANTS

In some railway applications, and for heavy machinery where oil of high viscosity at low temperatures was needed, the Russian oils were generally considered to be superior, whilst in engines, textile machinery and ordinary factory use the American products, oils which retained their viscosity at elevated temperatures, were preferred. Variations in the properties of oils were recognized and acknowledged to be of the utmost importance for, as Professor Thurston commented, the use of unsuitable lubricants often led to journals being welded into their bearings, the breakage of wagon and carriage axles, and the 'consequent destruction of trains loaded with passengers'.[49] Since mineral oils were cheaper than animal and vegetable oils there was an immediate economic incentive to their adoption. They possessed other advantages too. Unlike the animal and vegetable oils they do not contain acids nor do they experience the same tendency to decompose when exposed to heat, or due to age, which makes them particularly suitable for use in bearings where the presence of acids could lead to attack and corrosion of the bearing metals. Oxidation is also less, and this is a most desirable feature in a lubricant otherwise the oil thickens and becomes 'gummed' which increases the power required to drive the machinery.

The effect of heat on oils was of paramount importance and attracted the attention of numerous investigators, many of whom were directly concerned with either the textile industry or the railways. Absorption by cotton and other textile wastes of oils with a tendency to oxidize rendered them susceptible to spontaneous ignition due to the generation of heat during this chemical process, a process accelerated by the warm ambient temperature of the mills themselves, or, perhaps, the proximity of hot-water or steam pipes. Many conflagrations in textile mills were attributed to this cause and aroused the interest of insurance companies who sponsored tests on oils. The vegetable oils were especially dangerous in this respect as shown by tests conducted by Messrs Galletly and Coleman, and also by Professor Ordway. Whereas cotton waste soaked with boiled linseed oil would spontaneously ignite, under favourable conditions, in one and a quarter hours, the addition of mineral oils was found to have a profound effect and in the same circumstances combustion had not occurred after twenty-six hours.[50] However, this was not a precaution that could be taken indiscriminately because certain mixtures actually advanced ignition times. An early indication of the advantageous potentialities of additives in lubricating oils also occurred on the railway. In the same year that W. Bridges Adams presented his paper to the Institution of Mechanical Engineers another paper, describing the addition of lead oxides and india-rubber to whale oil, a composition which had been patented by Mr Donlan in 1848, was presented to the Institution by Mr John Lea who reported on experiments with this compound on express locomotives working on the Manchester and Crewe Section of the London and North Western Railway. Although the oil and tallow normally used for this purpose cost nearly 'fourpence per journal, per 1000 miles' that of the new compound was 'scarcely more than one penny for the same work.'[51] Whilst the value of mineral and compound oils was recognized quickly, an immense experimental effort was needed, and expended, to determine the most appropriate compositions for specific purposes; and so opened another avenue of investigation and experimentation which has led to the situation, today, something over a hundred years later, where few pure oils are used as lubricants and the role of the oil is largely that of a carrier for the variety of chemical substances which alter and enhance the function of the lubricant.

Though it was generally acknowledged that the best means of ascertaining the efficiency of a lubricant was to subject it to the actual conditions it would endure in service it was not always convenient or possible to do this. The alternative was to employ testing machines in which actual operating, or even more severe, conditions could be reproduced. A large number of such machines were designed and built ranging from those to suit specific requirements, such as Mr W. Stroudley's machine for testing railway journals and Boult's 'Cylinder Lubricant' Tester, to more general devices such as Professor Thurston's machine, mentioned earlier, and Ingram and Stapfer's Patent oil tester. Special versions of these general testing machines were produced, however, for the benefit of the railway companies. The capital outlay involved and the cost of the power to drive them was a good investment on the part of the large consumers because it could lead to substantial reductions, between 10% and 30%, in their expenditure on lubricating oils. Small users, who could not warrant the expense of a machine of their own, relied upon standard data and their own records to guide them in their selection of lubricants.

A variety of tests to determine the properties of oils — viscosity. acidity and decomposition, volatility, flash and firing points, solidification, gumming and spontaneous ignition, etc. — were established during the last quarter of the nineteenth century. These ranged from the extremely simple to those of a more complex nature requiring fairly sophisticated instruments. Thus, Thurston recommended that to detect the presence of grit in oils, a most undesirable inclusion since it caused undue wear, a drop of oil should be placed on clean white blotting paper which would absorb the oil while leaving the impurities as visible black specks on the surface.

The measurement of viscosity, or 'the body' of the oil as it was frequently called, also occupied a number of workers and clearly illustrates how the immediate problems were concentrating the endeavours of investigators throughout the world and leading them, in some instances, to identical solutions. In this case, the idea of permitting the oil to pass through a small orifice and measuring the time taken for some standard quantity to do so obviously occurred to several men simultaneously. Mr T.J. Pullin, who was associated with Mr J. Veitch Wilson in tests to compare animal and vegetable oils with compound oils,[52] invented such a machine as did Mr W.H. Hatcher whose design was subsequently improved by a number of people, most notably, perhaps, by Mr Boverton Redwood.[53] At this same period, the mid-1880s, in the U.S.A. a similar instrument was invented by Mr G.M. Saybolt, who was then an Inspector to the Standard Oil Company of New York while the European version was produced by Carl Engler. Papers on viscosity were published by Engler in 1885 and by Redwood in 1886.[54] The Engler, Redwood and Saybolt viscometers established themselves as the standard general industrial instruments and are still in use, each largely confined to the area of its origins. Although the viscosity is expressed in arbitrary units — Engler, Redwood, or Saybolt seconds — which do not relate simply to absolute viscosity, these instruments nevertheless provide a quick and easy method of checking viscosity at predetermined temperatures which no doubt accounts for their long and continuing service.

By the end of the nineteenth century, then, the mineral oil industry was well established not only as the major source of illuminating oils but also as the chief supplier of lubricating oils. Indeed, it is no exaggeration to claim that it was the constantly growing demand for lubricants that led to the development of the industry on the scale it achieved although, of course, it was given further impetus towards the very end of the century by the requirements of the internal combustion engine. Concurrently, techniques for assessing and measuring the properties of oils and producing compound lubricants to satisfy specific criteria were perfected.

The increasing population in Europe and the U.S.A., together with a rising standard of living expressed through greater demands for industrial products, especially textiles, plus the growth of a direct consumer demand for lubricants for bicycles, sewing machines and other domestic appliances would inevitably have led to an increase in output from the refineries. Unquestionably, the phenomenal expansion of the railways was a significant, probably the most significant, cause for the vast growth of output that actually happened.

Bearing Metals

EARLY ALLOYS

From the beginning of the railway era the design of axle bearings was closely related to the metals used. It has already been mentioned that early experience had led to half-bearings of cast iron situated above the axle, cast iron showing a smaller friction loss than bearings of brass, but this necessitated bearings at least four inches long to prevent them cutting into the axles, which deflected when the wagons were loaded. Similar problems were encountered with the tin bronzes which, lacking the plasticity necessary to cater for faulty adjustment, led to hot axle-boxes.

Isaac Babbitt invented a journal box which went a long way towards solving this particular problem. This was granted United States patent No. 1252 on 17 July 1839 and in the following year, on 15 May 1840, received British patent No. 9724; subsequently, in 1847, it was patented in Russia too.[55] His design included a soft bearing metal housed in a harder and stronger shell and he thereby obtained a greater load-carrying capacity without loss of plasticity which permitted the bearing surface to adjust itself to the contour of the shaft, or to accommodate a slight lack of alignment. His patent specifications included an incidental suggestion that a good bearing lining could be obtained with a ternary alloy consisting of 50 parts of tin, 5 of antimony and 1 of copper; he did not patent a specific composition but rather the idea of using a certain class of material as a lining metal. It so happened that this particular alloy, 89.3% tin, 8.9% antimony and 1.8% copper proved to be so successful and was used so extensively that Babbitt's name became associated with it and has continued to be so until the present. Indeed, his name is now given to the range of tin base alloys containing 3.5%–15% antimony, and his connection with the journal bearing is all but forgotten. But in its day it was very successful and earned Babbitt an award of $20,000 granted by Congress in 1842.

In Europe railway engineers were busily engaged with the problems of bearing design and within a few years a large selection of bearing metals had been tried. At the Nuremberg Railway carriage works tests were carried out, under the direction of F.A. von Pauli, with thirteen different bearing metals and the results, which were presented in 1849, constituted the first work ever published on journal bearings.[56] The most suitable alloy for this purpose was still being sought by the German railways in 1861–2 and the results of tests carried out at Göttingen and Hanover were published in a paper by Heinrich Kirchweger, printed in a German railway journal. The influence of British railway engineering at this time is clearly indicated by Kirchweger's use of British units.[57] Elsewhere similar work was being undertaken.

PHOSPHOR BRONZE

Interest was being shown in the use of phosphorus as an alloying element and as early as 1848–9 two patents covering the use of phosphorus in copper and brass had been taken out by Alexander Parkes of Birmingham (see the paper by Robert Friedel in this volume) but he did not mention bronze.[58] The value of phosphorus as an alloying element had also been predicted, about 1865, by Dr. J. Percy who was then the Lecturer in Metallurgy at the Royal School of Naval Architecture.[59] The bronzes were a promising field for exploration at this period because they seemed to offer the possibility of the desirable combination of hardness to withstand wear and sufficient ductility to bed to the shaft or journal. Here was a chance to use unlined shells. Although phosphor bronze was amongst the earliest of alloys to be studied in detail, the first patent for it under that specific name was not granted until 1870.[60] But by that time it was widely appreciated that the addition of small quanities of phosphorus produced significant improvements in the properties of bronze and experiments were being conducted from one end of Europe to the other. Large-scale tests were made, during 1870–1, on behalf of the Belgian Government and even earlier investigations had been made in Russia.[61] The discovery that the addition of phosphorus to bronze enhanced its bearing properties encouraged other experiments and so bronze metallurgy at this period became characterized by a 'many recipes' approach and almost anything that could be added to the 'stock pot' was tried — aluminium, iron, manganese, nickel, and other metals. It was also a time of innumerable patents many of which could hardly have been deserved. For example, although it had been common knowledge for some time that copper was hardened by the addition of small amounts of iron and manganese, Parsons sought, and was granted, a patent covering the addition of these metals in 1876.[62] Perhaps the memory of the $20,000 granted to Babbitt was still fresh and so chances could not be taken.

LEAD ALLOYS

It was the addition of lead to bronze which turned out to be of greater importance in bearing applications although the production of these alloys proved to be difficult and a lot of careful investigation was necessary before satisfactory methods were obtained. In England, about 1870, Alexander Dick produced a successful bronze with the composition 80% copper, 10% lead and 10% tin which was adopted for railway bearings and remained the standard bearing metal for many years.[63] The same composition was used in the United States, on the Pennsylvania railway, and it was from this company that the next advance came. Dr. C.B. Dudley, their chief chemist, published in 1892 the results of a long series of tests on actual bearings in service.[64] His findings revealed that the rate of wear and the tendency to overheat diminished with the increase of lead content, and increased as the tin content increased. His ternary alloy, Ex. B metal, with a composition 77% copper, 15% lead, and 8% tin, which became the standard bearing metal on his company's railways, represented the ultimate in his attempts to increase, satisfactorily, the lead content. Above 15% lead he could not prevent lead segregation. Seemingly, in his investigations, although he made alloys with a higher tin content he did not try reducing the tin. Precisely this modification enabled G.H. Clamer and J.G. Hendrickson to succeed. They found that the lead content could be safely raised to 30% if the tin was reduced to 5% and in 1900 they were granted a patent for a series of lead bronzes containing over 20% lead and less than 7% tin.[65] Their achievement may be explained by the fact that the reduction of tin reduced the amount of copper-tin eutectic[66] present which in turn accelerated the solidification and thus trapped the lead before it could sink through the partially solid alloy. The successful production of components in these alloys depended upon more than chemical composition — foundry skills and technique were of equal importance and an inability to achieve the necessary rapid cooling prevented these extra-high-lead alloys finding extensive use. Clamer, himself, found a variation of as much as 2% in the lead content between the top and bottom of large castings produced from the correct alloys, handled with the utmost care, under the very best conditions.[67] There were other ways to make high-lead bronzes and one of them by the addition of sulphur — which diminished the temperature range in which molten lead and copper are immiscible — discovered by A. Allan, led to a patent controversy between Clamer and Allan. Lead segregation is also reduced by the addition of other elements, notably nickel, silicon, zinc and zirconium.

Throughout the nineteenth century, from the appearance of Babbitt's lined bearing, the requirements of the rapidly spreading railways stimulated the search for better bearing metals and led to countless different alloys appearing on the market. Many of them were unnecessarily complex. Thus, Miller's lead base alloy which was marketed in the United States, in 1888, by Singley under the name of 'Magnolia' metal, included in its patent specification provision for the inclusion of aluminium, bismuth and silver, as well as tin. However, since bismuth and silver were common impurities in lead at that time it seems probable that their presence in the alloy was accidental, and attributable to the use of impure lead in the first instance by Miller, rather than as essential constituents of the resulting bearing alloy. Nevertheless, significant advantages were claimed for the metal, amongst them its adaptability for both light and heavy work. 'It is equally suitable for the heaviest plate-mill roll bearings and the lightest spindles of a spinning frame.'[68] Another advantage claimed for this metal was 'its capability of running for a considerable time without lubrication'. Now, of course, the main feature of all these successful bearing alloys was the presence of a low melting point constituent, mainly lead or tin, which in the case of excessive heating melted locally and formed a molten film which was smeared over the bearing surface and thus prevented seizure. It was this property that ensured the satisfactory performance of the alloy and many of the variations in composition that found their way on to the market contributed little, or nothing, to the metal's performance. However, this splendid spectrum of compositions was also reflected in the variety of names given to the alloys, names which ranged from the exotic — Camelia metal, Magnolia metal, Damascus bronze — to the mundane and utilitarian — anti-friction metal, car-box metal, antimonial lead. This situation led to intensive testing and the attempt to relate the analysis of the metals to their service behaviour. The programme at the laboratories of the Pennsylvania Railroad at Altoona led to the conclusion that some of the constituents present in very small quantities were only impurities and others, claimed as essential constituents, were of doubtful value. Although as the report cautiously admitted, it was possible that the experiments 'had not gone far enough to prove their value'. The predicament of both metal producers and users was neatly summarized by Thurston at the turn of the century, 'Because of the immense number of possible combinations of metal in alloys, and the great influence of slight changes of composition, the determination of the best bearing-metal is a difficult matter'.[69] Notwithstanding the difficulties, by this time certain criteria had been established for a good bearing metal:

1. It had to be strong enough to carry the load without distortion, which in the case of railway wagon journals meant sustaining pressures as high as 350–400 lbf/in^2.

2. The metal should not heat readily. Research had shown that, in general, the harder the bearing metal the more likely it was to heat.

3. The alloy should be amenable to successful manipulation in the foundry. Oxidation, which caused spongy castings, could be prevented by the addition of 1–2% of zinc, or by a small amount of phosphorus.

4. The metal should show little friction. Although it was realized that the friction was almost wholly a question of the lubricant used it was also recognized that the metal of the bearing exerted some influence as well.

5. The best bearing metal, other things being equal, is the one with the slowest rate of wear.

BEARING METALS IN THE TWENTIETH CENTURY

The development of the internal-combustion engine shifted interest to the tin base alloys which were much more suitable than those with a lead base where comparatively light loads and high speeds are involved. Normally they contained antimony to increase their hardness and copper to prevent segregation. But as with the bronzes, discussed earlier, cooling rates were critical for the production of sound castings and a lot depended upon the skill of the foundrymen in determining the correct pouring temperature which varied with the wall thickness of the casting and the temperature and mass of the mould. Continued progress in the design and construction of these engines, and in particular the requirements of aero-engines, called for metals with the capacity to withstand higher pressures than the ordinary tin base alloys. A range of copper lead alloys, containing, a high proportion of lead, was found to satisfy these needs, and the alloys proved to be extremely successful for heavy-duty applications in aircraft engines and in the bearings of Diesel engines. In tests on heavy lorries some of these bearings completed 100,000 miles in service.[70] These lead bronzes are to be distinguished from those mentioned above, and used in railway work, by the almost complete absence of tin. Since lead is not soluble in copper the value of these alloys as bearing metals depends upon securing a proper dispersal of the lead through the copper matrix. This requires a rapid cooling rate but the prevention of lead segregation is also assisted by the addition of very small amounts of tin (about 0.07%) and nickel (about 0.01%).

For quite another reason these bronzes were of considerable interest. Their structure did not conform to the generally accepted ideas about bearing metals and, in the event, contributed to the development of the theory of these materials. The fact that Babbitt's alloys consisted of hard particles scattered throughout a soft matrix combined with the indisputable fact that they also proved to be the best bearing metals then available established the theory that a good bearing metal must possess this structure. Innumerable scientific investigations throughout the entire century did not seriously challenge this idea which has persisted almost to the present day.[71] However, doubts were entertained about the correctness of this view in the mid-1930s when Bassett wrote of two schools of thought on the subject, one holding that only a two-phase metal, a hard constituent in a soft matrix, would serve as a bearing metal, the other maintaining that a single-phase metal would work perfectly well.[72] Bassett himself obviously inclined to the former view and attributes the unsuitability of the silver-copper-cadmium alloys as bearing metals to their lack of hard crystals.[73] Rather interestingly, when discussing the copper lead alloys he states that their value as a bearing metal 'depends upon the efficiency with which the lead is dispersed through the copper matrix'[74] but he makes no further comment on the structure. Here the commonly accepted structural form is inverted and the matrix is the hard phase while the soft phase (lead) is embedded within it. Seemingly, preoccupation with the attempt to produce alloys consisting of hard micro-constituents in a softer matrix deflected attention away from other useful forms of bearing metal for many years and it is only fairly recently that two other categories have been established, those in which a softer metal is distributed through a hard matrix and the single metal or single-phase alloy. During the Second World War, with its demand for higher performance in military aircraft, silver bearings were used to carry the heavy loads and were of remarkable value. These were plated with lead, to reduce the risk of seizure, which in turn was coated with indium to increase the corrosion resistance of lead to acidic oil.[75]

Over the years, then, first to meet the demands of the railways and then in response to the requirements of internal combustion engines, an immense effort has been applied to the production of bearing metals and through a process of trial and error, of failure and success, a variety of excellent alloys have been obtained. But despite its success in the past it seems that this approach must be reaching its limit.

Summary

The rapid development of the railways which began about 1830 and gathered momentum from about 1850 created a number of problems connected with journal bearings. By prodigious labour, conscientiously and patiently carried out, successful solutions were found although the reasons for the success were not always appreciated at the time, or for many years afterwards in some cases. Here, W. Bridges Adams' axle-box, which was used for some thirty years before Beauchamp Tower discovered hydrodynamic lubrication, springs to mind.

Whether in the design of bearings, the mixture and testing of lubricants, or the production of bearing metals, it was through painstaking, methodical, empirical research that progress was made. This intense, persistent technological activity stimulated scientific work which, in turn, produced a substantial growth in theoretical knowledge and at times culminated in some high peaks of understanding. The unconnected, but simultaneous, work of Reynolds and Petroff is indicative of the mounting social, economic and technical pressures guiding the direction of scientific research. It appears to have been a one-way process. In no major sphere can one discern an instance of theory leading practice.

By its unquenchable thirst for lubricants the expanding railway system contributed significantly and progressively to the growth of the petroleum industry which became, by far, the biggest supplier of lubricants. The mineral oils, almost invariably compounded, were available in adequate quantities, they were superior in performance and they were approximately only a quarter of the cost of the lubricants they replaced. Furthermore, this transition permitted the release of animal and vegetable fats and oils for human consumption which, in itself, was a great benefit to the growing populations of the industrial countries.

At the dawn of the present century — which has seen a momentous exploitation of the internal combustion engine, comparable with that of the steam engine in the nineteenth century — a vast quantity of relevant experience and knowledge was available which constituted a vital part of the foundation of this exploitation. And so, too, was a firmly established petroleum industry ready and able to supply the necessary fuel cheaply. In a not insignificant manner, then, did the needs and achievements of the railway contribute to the success of the internal combustion engine.

Notes

The author would like to thank the Librarian of the Institution of Mechanical Engineers for his help and his permission to reproduce Figures 1, 3 and 4 from the *Proceedings* of the Institution.

1. The Lake Nemi ship in Italy, believed to have been constructed about AD 40, contained the remains of a bearing running on bronze balls. The sketch books of Leonardo da Vinci and the drawings of Ramelli contain a number of depictions of anti-friction rollers.
2. See, for example, the illustrations in Book VI of Agricola's *De re metallica*, Basel 1556.
3. Plumier, Charles *L'Art de Tourner en Perfection*, Lyon 1701, Plate 104.
4. Dowson, D. *Lubricants and Lubrication in the Nineteenth Century*, The Newcomen Society, London 1974, p. 3.
5. Smith, N. *Man and Water*, London 1976, Ch. 12.
6. Fairbairn, W. *Treatise on Mills and Millwork* Part II, London 1863, pp. 212, 242, etc.
7. *ibid*. p. 72.
8. Thurston, R.H. *Friction and Lubrication*, London 1879, pp. 26–7.
9. *ibid*. pp. 38–9.

10. *ibid*. pp. 38–9.

11. *ibid*. p. 39.

12. *ibid*. p. 40.

13. Oeynhausen, C. von and Dechen, H. von *Railways in England, 1826 and 1827*, The Newcomen Society, London 1971, p. 29.

14. *ibid*. p. 30.

15. Adams, W. Bridges 'On Railway Axle Lubrication' *Proc. I. Mech. E.* London 1853, p. 59.

16. Wheeler, G.U. *Friction and its Reduction*, London 1903, p. 21.

17. Thurston, R.H. *Friction and Lubrication*, p. 31.

18. Wheeler, G.U. *op. cit*.p. 34.

19. Thurston, R.H. *op. cit*. pp. 128–9.

20. Thurston, R.H. *op. cit*. p. 129.

21. Wheeler, G.U. *op. cit*. p. 89.

22. Thurston, R.H. *op. cit*. Table p. 13.

23. Thurston, R.H. *op. cit*. p. 209.

24. Tower, Beauchamp 'First Report on Friction Experiments' *Proc. I. Mech. E.* 1883; 'Second Report on Friction Experiments' *Proc. I. Mech. E.* 1885.

25. *Proc. I. Mech. E.* Nov. 1883, pp. 653–9; Jan. 1884, pp. 29–35.

26. Tower, Beauchamp 'First Report' *Proc. I. Mech. E.* Nov. 1883, p. 634.

27. *ibid*. p. 635.

28. *The Engineer*, 29 Feb. 1884, p. 164.

29. Reynolds, Osborne 'On the theory of lubrication and its application to Mr. Beauchamp Tower's experiments' *Phil. Trans.* 177(1) 1886, pp. 157–234.

30. Rankine, W.J.M. *A Manual of Machinery and Millwork*, London 1869, p. 18.

31. Thurston, R.H. *Friction and Lubrication*, p. 212.

32. Thurston, R.H. *Friction and Lost Work in Machinery and Mill Work*, New York 1887, p. 393.

33. Cameron, A. *Principles of Lubrication*, London 1966, p. 272.

34. Cameron, A. *op. cit*. p. 267.

35. *Proc. I. Mech. E.* Jan. 1884, p. 34.

36. *Proc. I. Mech. E.* Vol. 161, 1949, p. 71.

37. Kingsbury, A. 'On problems in the theory of film lubrication with an experimental method of solution' *Trans. A.S.M.E.* Vol. 53, 1931, p. 59.

38. Needs, S.J. 'Effects of side leakage in 120 degree centrally supported journal bearings' *Trans. A.S.M.E.*: Vol. 56, 1934, p. 721; Vol. 57, 1935, p. 135.

39. Ocvirk, F.W. 'Short bearing approximation for full journal bearings' *N.A.C.A.*, T.N. 2808, 1952. Dubois, G.B. and Ocvirk, F.W. 'Experimental investigation of eccentricity ratio, friction and oil flow of short journal bearings' *N.A.C.A.*, T.N. 2809, 1952. Dubois, G.B. and Ocvirk, F.W. 'Analytical Derivation and Experimental Evaluation of Short Bearing Approximations for Full Journal Bearings' *N.A.C.A.* Rep. 1157, 1953.

40. 1858 Patent No. 42 was granted to J.A.M. Chafour for a rolling bearing axle box. 1862 Patent No. 240 was granted to W.E Newton for a ball bearing for railway rolling stock. 1873 Georg Wieckum, of Budapest, was granted Patent No. 1685 for a design which included both ball and roller bearings.

41. Wilson, S.S. 'Bicycle Technology' *Scientific Technology and Social Change, Readings from Scientific American*, San Francisco, 1974, p. 62.

42. British Timken Ltd. *Timken Tapered Roller Bearings Applied to Railway Rolling Stock and Outdoor Machinery*, Birmingham 1933, p. 5.

43. Dowson, D. *Lubricants and Lubrication in the Nineteenth Century*, The Newcomen Society, London 1974.

44. Williamson, Harold F. and Daum, Arnold R. *The American Petroleum Industry: 1858–1899; The Age of Illumination*, Evanston, 1959. Appendix Table D: 1, p. 742.

45. The figures given in this table are based on extracts from *The Fontana Economic History of Europe. The Emergence of Industrial Societies — 2*. Editor Carlo M. Cipolla. London, 1973. Railways. Table 1, p. 789.

46. Williamson, Harold F. and Daum, Arnold R. *op. cit*. p. 679.

47. *ibid*. p. 742.
48. *ibid*. p. 645.
49. Thurston, R.H. *Friction and Lubrication*, London 1879, p. 43.
50. Wheeler, G.U. *Friction and its Reduction*, London 1903, pp. 61–2.
51. Lea, John 'On a New Lubricating Material' *Proc. I. Mech. E.*, London 1853, p. 67.
52. Wheeler, G.U. *op. cit.* p. 69. The results of these tests, published in *Engineering*, 20 April 1888, p. 379 showed that properly prepared compound oils offered less frictional resistance and ran cooler than any simple animal or vegetable oils of similar gravity and viscosity.
53. *ibid*. p. 79.
54. Cameron, A. *Principles of Lubrication*, London 1966, footnotes, p. 20.
55. Details on Isaac Babbitt derived from *Dictionary of American Biography*, Vol. 1, p. 456.
56. Cameron, A. *op. cit.* p. 263.
57. *ibid*. p. 264.
58. Bassett, H.N. *Bearing Metals and Alloys*, London 1937, p. 5.
59. *ibid*. p. 5.
60. *ibid*. p. 6.
61. *ibid*. p. 6.
62. *ibid*. p. 6.
63. Corse, W.M. *Bearing Metals and Bearings*, New York 1930, p. 16.
64. *ibid*. p. 16.
65. *ibid*. p. 16.
66. 'Eutectic' describes a laminated type of microstructure formed during solidification by two metals which are completely soluble in the liquid state but completely insoluble in the solid state, and they solidify by crystallizing out as alternate layers of the two pure metals.
67. Bassett, H.N. *Bearing Metals and Alloys*, p. 325.
68. Wheeler, G.U. *Friction and its Reduction*, London 1903, p. 135.
69. Thurston, R.H. *Friction and Lost Work in Machinery and Mill Work*, p. 409.
70. Bassett, H.N. *op. cit.* p. 321.
71. Higgins, R.A. *Engineering Metallurgy*, Part I, London 1968, p. 366.
72. Bassett, H.N. *op. cit.* p. 80.
73. *ibid*. p. 364.
74. *ibid*. p. 321.
75. Rollason, E.C. *Metallurgy for Engineers*, London 1961, p. 337.

Bibliography

Bassett, H.N. *Bearing Metals and Alloys*, Edward Arnold, 1937.
Cameron, A. *Principles of Lubrication*, Longmans, 1966.
Corse, W.M. *Bearing Metals and Bearings*, The Chemical Catalog Co., Inc., 1930.
Fairbairn, W. *Treatise on Mills and Millwork*, Parts I and II, Longman, Green, 1863.
Higgins, R.A. *Engineering Metallurgy*, Part I, The English Universities Press, 1968.
Oeynhausen, C. von and Dechen, H. von, *Railways in England, 1826 and 1827*, The Newcomen Society, 1971.
Rankine, W.J.M. A Manual of Machinery and Millwork, Charles Griffin & Co., 1869.
Rollason, E.C. *Metallurgy for Engineers*, Edward Arnold, 1961.
Smith, N.A.F. *Man and Water*, Peter Davies, 1976.
Thurston, R.H. *Friction and Lubrication*, Trübner and Co., 1879.
—— *A Treatise on Friction and Lost Work in Machinery and Millwork*, John Wiley, 1903.
Wheeler, G.U. *Friction and its Reduction*, Whittaker & Co., 1903.
Williamson, Harold F. and Daum, Arnold R. *The American Petroleum Industry: 1859–1899; The Age of Illumination*, Northwestern University Press, 1959.

The Waltham Pound Lock

K.R. FAIRCLOUGH

In 1571 an Act of Parliament was passed 'for the brynging of the Ryver of Lee to the Northside of ye Citie of London'.[1] This act, sponsored by the city authorities, gave details of plans proposed by the aldermen for building a new cut from the River Lee, through Hackney and Shoreditch, to terminate just outside the city walls near Moorgate. Once this canal had been completed, the aldermen further proposed that improvements be carried out along the existing river between Ware and the mouth of their new cut.

These ambitious plans, however, were never carried out. During the committee stage of the bill's passage through the House of Commons, several important additions and alterations were made to the original draft, the most important of which was the insertion of a veto on the right to collect tolls for using either the improved river or the new cut. Such a veto meant that the City was unable to finance the canal and although the scheme was not shelved immediately it became obvious within a couple of years that the project could not go ahead.

Interest in improving the Lee had been awakened, however, and on 27 September 1574 a Commission of Sewers was appointed to improve the navigation on the river[2] in order that it might become an important artery along which grain, meal and malt could be brought to the capital. John Norden ascribes this later initiative to 'the instant suyte of the inhabitants of Hartfordshire'[3] who saw that an improved navigation would allow them to capture an important share of a trade which had previously been dominated by land carriers known as 'badgers'.

There is no evidence to suggest that the Commissioners ever seriously considered building the new cut to Moorgate; instead they concentrated on improving the existing navigable channel. They scoured and cleaned this channel, ordered the removal of all fishing weirs and impediments to navigation, raised all bridges over the river to allow the barges more headroom, laid out a rough towpath which included towing bridges, and came to a series of differing compromises with the many millers in the valley whereby they were to take water out of the river without the use of any flash locks which would hold up the passage of the barges.[4] Though these improvements were much less ambitious than those originally proposed by the City, they were nevertheless extremely successful. Within four or five years of the Commissioners' appointment the river had been substantially improved and an expanding barge traffic was already arousing the opposition of the badgers.

The most ambitious task undertaken by the Commissioners was the construction of a pound lock at Waltham and it remains their best-known work for it was the first pound lock in England to be equipped with mitre gates at both ends. The only earlier pound locks known in England were those built between 1564 and 1567 by John Trew along the River Exe. These were constructed as large pools in which several boats could lie at once. Mitre gates were used at one end only; at the other end were single guillotine gates.[5]

The Waltham lock is described in a poem written by William Vallans some time during the 1580s.[6] In it two swans make a journey down the River Lee and one of the many sights they marvel at is:

> But newly made, a waterwourke: the locke
> Through which the boates of Ware doe passe with malt.
> This locke contains two double doores of wood,
> Within the same a Cesterne all of Plancke,
> Which onely fils when boates come there to passe
> By opening of these mightie dores with sleight,
> And strange devise, but now decayed sore.

Before the Commissioners began their work barges navigated the river above Waltham by means of a flash lock which stood across the main stream about three-quarters of a mile above Waltham High Bridge. When shut, this flash lock diverted water out of the river into the head stream of Waltham Mill. Since few barges navigated the river at this time the flash lock was usually shut and consequently, it was later claimed, 'the auntient Channell did decay for lacke of Continuall corse of Water and soe did become unpassable for Boates'.[7]

The Commissioners, however, chose not to scour and cleanse this traditional channel, but rather to open a new route. They ordered that 'the passage of the Boates shold be directed to passe by the sayde Mill-streame And for that purpose that a newe Cutt shold be made from the sayde Millstreame somewhat distaunte from the sayde Mill towardes the North unto the old River towardes the West', and that along this new cut 'a newe devised Lock to Cawse the water to swell upp wherby Boates may passe and repasse betwixt the sayd river of Ley and the water belonginge to the Mill' be constructed.[8]

To complete the arrangements for this new route the Commissioners further decreed that the old flash lock be pulled up and replaced by a 'loweshare of three foote and a half highe from the bottome of the river for the forcinge of the water to his (Edwards Denny's) sayde Milles and yett not be suffered to be anie higher for that all white and superfluous waters may passe that way for the drayninge of the groundes adioyninge and keepinge open of the olde channell'.[9] As a precaution the Commissioners added that the bargemen were to have the right of pulling up this loweshare and using the traditional channel, if, for any reason, they were unable to use the newly opened route. This particular proviso was to assume great importance later when (in 1592) disputes arose and riots ensued over the rights of navigation through Waltham.

In October 1576 the Commissioners were still debating which route to take through Waltham.[10] Yet by October 1577 they were meeting to discuss the completion of their work in the area. Thus the pound lock must have been built some time during the spring or early summer of 1577 and not earlier as is usually stated.

Besides discussions about the route the Commissioners seem to have considered at least two alternative plans for the construction of the lock. A surviving estimate of costs,[11] reproduced in full in the Appendix, provides a rather inadequate comparison between the cost of building the lock entirely of wood, and an alternative proposal whereby the walls of the pound would be built of stone. Vallans' poem suggests that the Commissioners chose to build entirely with wood but no firm evidence remains to throw light on the reasons for such a choice.

The estimate itemizes the different sections of the lock and calculates the amount of timber necessary for each section. Although the arithmetic is not quite correct a total of 44 loades of timber is estimated and it was reckoned that the cost of this timber and the associated carpentry work would be £40 5s. 0d. No allowance, however, is made for any mechanism to open the doors nor for any paddles or other devices to let water into the lock when the gates were shut. The estimate for the lock built with stone walls is even more incomplete, although it does seem likely that this would have been the more expensive piece of work.

if it be don with timber	Tymb xliiij lodes at x^s	xxii^{li}		
	Sawinge + Carpenters woorke	xviij^{li}	v^s	
		xl^{li}	v^s	
the same with stone	Tymber xxiiij lodes	xii^{li}		
	Lyme viij lodes at 16.8	vi^{li}	xiii^s	iiij^d
	Sand xvj lodes at ii^s		xxxii^s	
	The walling	viij^{li}		
	Carpenters woork			
	Stone			
		xxviij^{li} [*sic*]		

In addition it was estimated that the digging of a new cut, 36 poles in length,[12] would cost a further £48, while heightening Waltham High Bridge by two feet would cost £4. These estimates, however, turned out to be much too optimistic. Though the total estimated cost of the work was given as £88 5s. 0d. beforehand, the actual cost was reported to be £271 18s. 0½.[13]

Assuming that the wooden lock was built according to the specifications laid down in the estimate, the reconstruction shown in Figures 1a and 1b indicates how the frame of the lock was constructed.[14] First, timbers known as 'grounsylles' and 'gysses' (joists) were laid in the bed of the cut to provide a foundation upon which to fix the frame of the

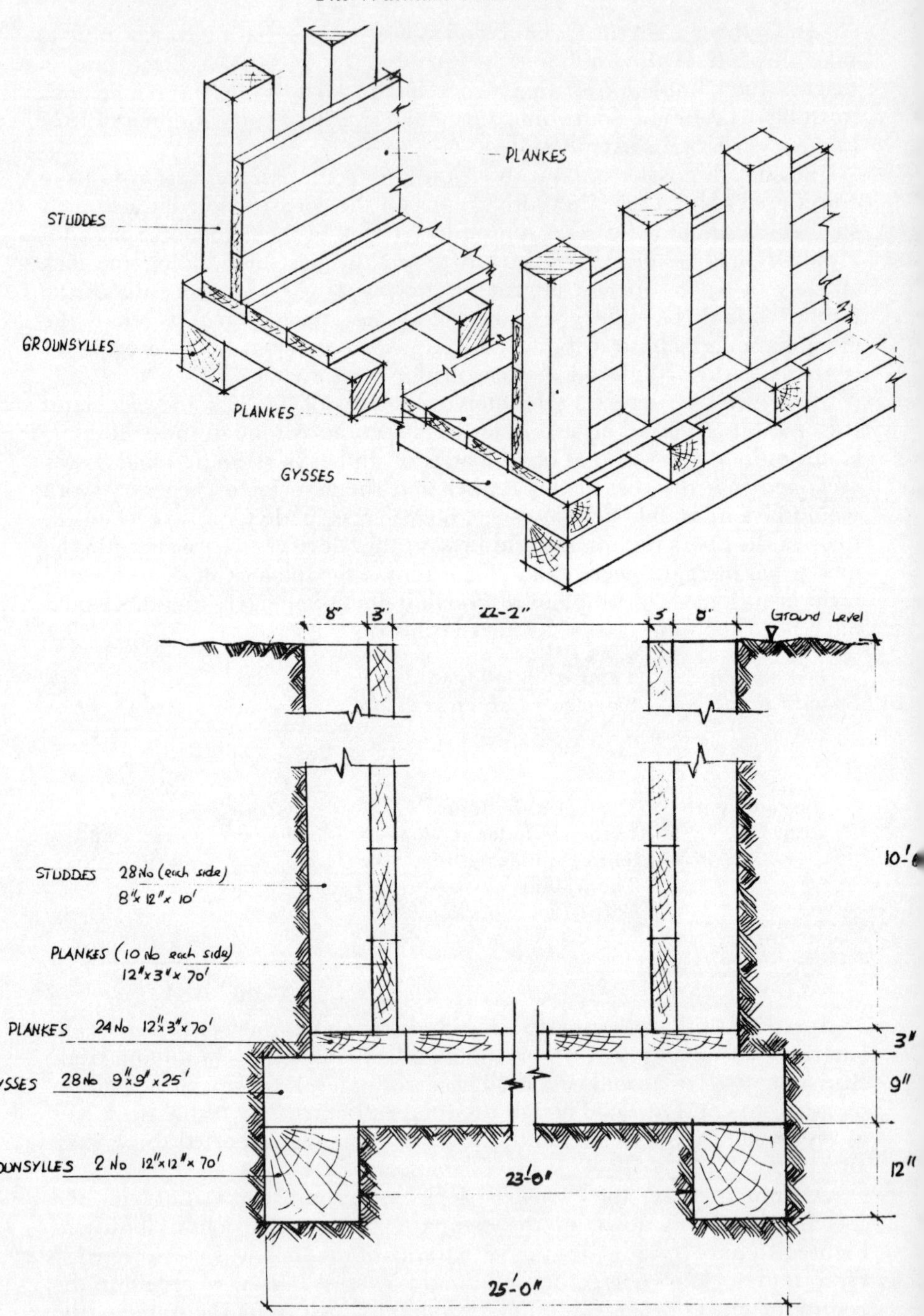

Figure 1. Pound Lock, Waltham Abbey, 1576.
(a) Isometric view.   (b) Central cross-section.

lock. The floor and walls of the enclosure were made up of planks, each measuring 70 feet by 1 foot by 3 inches, laid side by side. The floor planks were nailed to the 'gysses', whilst the wall planks were nailed to vertical 'studdes' which were fixed at regular intervals along the sides of the lock. In this manner a frame 70 feet long, 24 feet broad, and 10 feet high was built up, and at the ends of this frame posts were fixed from which the four gates were hung.

Vallans mentions that the gates were opened by a 'sleight and strange devise', but unfortunately the estimate gives no idea what this might have been. Conceivably chains were used because in 1581 it was alleged that 'Aron Yong of Waltham Abbey Tailor solde a chayne that belonged to the said lock to one Davie of Waltham Cross smyth for sixe pence'.[15] On the other hand such a chain might have been used to lock the gates so that they could not be opened except with the cooperation of the miller.

It seems likely that the Commissioners opened this new route after representations from the owners and occupiers of Waltham Mill for it was the owners who were made to bear the cost of the new cut and lock. The mill was owned by the Denny family as part of the manor of Waltham. At the time of the alterations Henry Denny had just died and his heir, Edward Denny, was still a minor and as such a Ward of Court whose interests were looked after by Queen Elizabeth. In view of this the Commissioners at their meeting in the Star Chamber decided that the cost of the work should be split equally between the executors of Henry Denny's will and the Queen in her role as Edward's guardian. All further costs of maintenance were to be borne by the miller, Thomas Perrott, during the remaining term of his lease after which they became the responsibility of Edward Denny and his heirs.

Since it was also reported that, despite their previous orders, the old flash lock was still standing, the Commissioners further decreed that the miller was to pull it down and build a loweshare by 14 February 1578 or else face a fine of £10.

Within a couple of years, however, major problems had to be faced. In 1579 full-scale repairs had to be carried out and the fact that the cost of this work was borne by the Treasury rather than the miller does suggest faults in the original construction. From hints in an account book[16] detailing these repairs it seems that one side of the wooden foundations had settled so far into the bed of the new cut that the lock had become lopsided. To rectify this the Commissioners were forced to close down the new cut so that the lock could be dismantled and masonry foundations constructed. The wooden frame was then replaced on top of these new foundations.

On the back of the account book a total cost of £96 13s. 3d. is recorded of which £40 6s. 9d. is attributed to 'redy money' and the remaining of £56 6s. 6d. to 'bills of Woork'. No further details are given about the 'redy money' but the account lays out in great detail the various charges which constituted the 'bills of Woork'. These include the wages paid, the amounts spent on transportation and sums for many small items such as nails and planks, the need for which must have arisen during the actual course of the reconstruction.

Work commenced on 18 May 1579 and was completed by 13 June. During the first few days temporary dams were built and preparations made so that five carpenters could be employed at 'pulling upp the plankes and laiing them agayne'. In between two masons and their assistants were employed to lay the masonry foundations. These carpenters and masons, being skilled workers, were paid 14d. per day.

Most of those employed, however, were unskilled men taken on either as general labourers or 'skavellmen'. The latter were so named because they were hired to bale out any water which seeped through the temporary dams using a tool known as a skavell. The two types of unskilled labour were interchangeable, men working at whatever job was available on any one day. Labourers worked during the day only while the skavellmen were needed right round the clock. Since the men were paid either 10d. or 12d. per shift it seems likely that they worked for either 10 or 12 hours and were paid at a rate of 1d. per hour, the variation in pay per shift not depending on the type of work done. Some men only worked the odd shift whilst others took advantage of the availability of employment to put in extremely long hours. One man, John Foster, earnt 9s. one week and 11s. the following, which means that in the second week he worked the equivalent of 7 day and 4 night shifts.

The account book also mentions that a Mr Trewe was paid £6 13s. 4d. for directing and overseeing the work. It is conceivable that this was the John Trew who was responsible for improving the River Exe. Such a high salary does suggest that he was being rewarded for professional skills over and above the mere ability to oversee a work-force. Whether he introduced any basic changes into the design or operation of the lock cannot be said.

The problems experienced with the Waltham lock arose no doubt because of the technical novelty of the work. The Commissioners, however, enjoyed far more success with their other improvements. Their success, however, only aroused opposition. Not only did the millers, the fishermen and the riparian landowners feel that their traditional rights had been interfered with but the local badgers felt that their livelihood was directly threatened. Traditionally grain and malt had been brought to markets in the lower Lee valley, particularly Hoddesdon, where local dealers and badgers bought it in order to resell in the London markets. This traditional pattern was threatened by a growing barge traffic, not only because barges could carry at cheaper rates, but also because markets further up the valley, particularly those at Ware and Hertford, could intercept much of the trade at the expense of Hoddesdon. The badgers, especially those living in Enfield, Cheshunt and Waltham, were to become the most persistent and vociferous opponents of the navigation.

In 1580 the badgers petitioned Lord Burghley requesting that the navigation be closed down. He advised them that he did not have the necessary authority and suggested that they approach Parliament instead. This they did but before their petition could be heard Parliament was prorogued.[17] Having failed by legal methods the badgers turned to violence and during the summer and autumn of 1581 destroyed many of the works along the river. A full-scale enquiry was held at which the badgers were given full opportunity to express their grievances. Nevertheless the Commissioners obtained the authority to carry on with their work without any of the badgers' objections being met. It was felt that the national interest warranted the expansion of barge traffic along the Lee even though it was admitted that the badgers did suffer as a result. For the rest of the decade the badgers continued to petition the authorities but to no avail and so they turned once more to violence. The outbreak in 1592 was better organized and barges were forced to stop using the river while appeals to the authorities to restore the navigation were made. Once more the verdict was in the bargemen's favour.

During both outbreaks of rioting Waltham Lock was a prime target. In May 1581 an approach was made to an employee at Cheshunt Mill for the loan of a handsaw with which to damage the lock but come July the less energetic course of setting the lock on fire was preferred. William Shanbrooke journeyed to London to purchase 'Rosseyn and brymston' for a groat. The lock was eventually fired in August but the damage was not as severe as hoped. One local inhabitant, Christopher Pennyfather, told bargemen who were using the lock several days later that he wished there had been a barrel of gunpowder in the lock when it had been set on fire.[18]

In 1592, however, the lock was completely destroyed.[19] In June of that year Edward Denny ordered his servants to dismantle the lock, to block up the new cut and to lower Waltham High Bridge to the height it had been prior to 1576. Denny argued that the Commissioners had made their alterations whilst he had still been a minor and therefore it could not be assumed that he had given his permission to changes affecting his freehold property and traditional rights. There were, however, veiled accusations that he had received money from the badgers to encourage him to take the steps he did. Whatever the case, however, his orders were quickly carried out.

Deprived of access to the new route the bargemen tried to take advantage of the provisions made by the Commissioners for just such a contingency. When, however, they tried to pull up the loweshare and proceed down the old river channel they met with violent opposition from large gangs of badgers and other local inhabitants who had gathered on the banks to thwart the bargemen's efforts. Throughout the summer, autumn and early winter, the violence continued. Boats were damaged, one was even sunk, and men on both sides suffered injuries. Eventually the bargemen found it impossible to continue, so once more they had to appeal to the central authorities for help.

In their case before the Star Chamber, however, the bargemen made no complaint whatsoever about the destruction of the pound lock; they concentrated solely on establishing their rights to use the original river channel. Indeed the destruction of the lock is only mentioned in passing by one of Edward Denny's servants under cross-examination. This suggests either that the bargemen felt their case to be stronger if they accepted Denny's arguments about his freehold property and concentrated on their own traditional rights, or that they were dissatisfied with the workings of the lock. Unfortunately there is no evidence to establish conclusively which was the case.

Though the bargemen succeeded in reopening the navigation the pound lock was never rebuilt. This was because the navigation reverted to the traditional route and the new cut in which the pound lock had stood was no longer used. The remaining evidence is incomplete[20] and on some points contradictory, but what does emerge is that soon after the end of the Star Chamber case some form of compromise was effected between Denny and the bargemen.

The evidence definitely establishes that Edward Denny built a turnpike, in fact a flash lock, at which he collected a toll of five shillings from each passing barge. This turnpike was still in existence in the eighteenth century despite many complaints from the bargemen that the lords of the manor of Waltham had no right to collect such a toll. The position of the eighteenth century turnpike on the map (Figure 2) is taken from a contemporary survey[21] but the turnpike which Denny built must have been roughly the same position, as must also the flash lock which was replaced by a loweshare in 1576–7. In evidence before a 1682 Commission of Sewers[22] certain local inhabitants claimed that Denny had built a new cut of over a mile in length in which the turnpike was sited but this seems very unlikely. It is much more probable that the old river channel was scoured and cleaned and that local memory had confused the measures taken by the Commissioners of Sewers in 1576–7 with those taken by Denny after 1595.

The confused evidence given to the 1682 Commission also mentions a 'Longe Poole very neere the said Corne Mills' through which barges had passed prior to the making of Denny's turnpike. P.J. Huggins[23] has already suggested that this 'Longe Poole' must have been the same as the cut built by the Tudor Commissioners and this is confirmed by measurements taken in the area. The new cut was said to have been 36 poles in length,[24] which would have made it just under 200 yards if it is assumed that 1 pole equals $16\frac{1}{2}$ feet. This approximates to the distance along the 'Longe Poole' as it is shown on the map in Figure 3. Any cut between the mill stream and the river beginning at a point any further up the mill stream would have been some 40–50 yards longer. It thus seems indisputable that the pound lock was situated at the place marked on the map.

There is no trace of this 'Longe Poole' today and it seems unlikely that

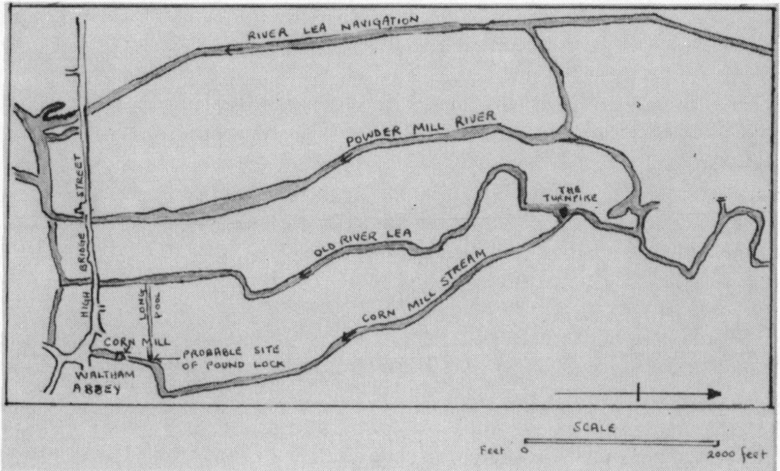

Figure 2. Branches of the River Lee near Waltham Abbey showing the site of the pound lock and of the turnpike. From a contemporary survey.

any remains of the lock would be found even if the site were excavated. However there is evidence that barges could still use the 'Longe Poole' until well into the eighteenth century. Notes in the margin of the 1682 Commission of Sewers' records mention that barges used the route within living memory though they had to carry much lighter loads than if they had used the turnpike.

Appendix

The estimate of costs to construct the pound lock at Waltham is taken from Public Record Office, State Papers Elizabeth, Domestic Series, Vol. 109 no. 133. A modern English version follows the original.

26 November 1576

> An estimate of the Charges of the newe Locke
> that shalbe made at Waltham yf yt be all of
> tymber as followeth, and beinge in leyngthe
> lxx foote and xxiiij foote in breathe

Imprimus xxviij gysses to lye in the bottom, under the plankes, every gysse being ix enches square and xxv foote longe so every gysse will Conteyne xvj foote, one quarter of a foote whiche will amownte to ix Lodes of square tymber.

Itm for the bottom xxiiij[ti] plankes, every planke Conteyninge lxx foote Longe, and one foote brode + iij enches thicke, so every planke will

Conteyne xvij foote &, whiche will amownte to viij Lodes xx foote of square tymber.

Itm ther, moste be ij grounsylles, every grounsyll of lxx foote Longe and one foote square, which will Conteyne vij^{xx} foote of square tymber, that is ij Lode, forty foote.

Itm for every syde ther moste be xxviij studdes, every studd beinge xij enches brode and viij enches thicke and x foote heygh, so every studd will Conteyne vii foote in square tymber, which will amownte to viij Lodes.

Item ther moste be ij peces of tymber to Lay upon the sayd studdes, every pece of lxx foote longe, and one foote square, which is ii Lodes xl foote.

Itm every syde will aske x plankes every planke beinge lxx foote Longe, one foote brode, and iij enches thicke, so every planke will Conteyne xvij foote &, whiche will amownte to, for bothe sydes, vij Lodes of tymber.

Itm the iiij gates wth the postes to hange them upon, and Certeyne Lande keyes for the same, will aske vi Lodes of tymber.

Suma of the Lodes of tymber xliiij   }
every Lode Rated at x^{s} the lode   } xxij^{li}

Itm the sawinge worke of all the foresayd tymber will amownte to lvj hundred at xx^{d} the hundred iiij^{li} xviij^{s} iiij^{d}

Itm the Carpenters worke for the same Locke will Coste xx markes so that the Carpenter be put to no other Charge but onlye the framynge, and settynge upe of the same.

Itm yf the sydes and endes of the sayd Locke be made of stone or brycke, then the tymber y^{t} shale go to the same will amownte to xxiiij Lodes or ther abowtes.

Itm the workmanshipe of the same walle, beinge x foote heyghe, and iij foote thicke will Coste xvj^{s} a pole, so that the mason be putt to no other Charge, but only the Reysinge of the same walle, and the same worke will Conteyne by estimacon x pole, every pole being xvj foote.

Itm ther will go to the same walle viij Lodes of Lyme whiche will Coste xvj^{s} viij^{d} the Lode, whiche is vi^{li} xiij^{s} iiij^{d}.

Itm ther moste be to the same worke xvj Lodes of sande at ii^{s} the Lode, xxxij ^{s}

Itm the heyghtinge of the hyghe brydge at Waltham to Reyse yt ii foote
hygher will Cost by estimacon iiij^{li}.

Item the Charges of the Cuttynge of the newe Cutte from the myll ryver to
the ryver of Lee, which Conteynes xxxvi pole, every pole rated at xxvj^{s}
viij^{d} the pole, whiche will amownte to xlviij^{li}.

if it be don	Tymb xliiij lodes at x^{s}	xxii ^{li}		
with timber	Sawinge + Carpenters woorke	xviij ^{li}	v^{s}	
		xl ^{li}	v^{s}	
the same with	Tymber xxiiij lodes	xii ^{li}		
stone	Lyme viij lodes at 16.8	vi ^{li}	xiii^{s}	iiij^{d}
	Sand xvj lodes at ii^{s}		xxxii^{s}	
	The walling	viij ^{li}		
	Carpenters woork			
	Stone			
		xxviij^{li} [*sic*]		
	The heightening of the bridge	iiij ^{li}		
	The chardge of the Cut	xlviij ^{li}		
		lij ^{li}		

MODERN VERSION
26 November 1576

> An estimate of the cost of the new lock
> that shall be made at Waltham if it be
> all of timber as follows, and being in
> length 70 feet and 24 feet in breadth

First of all 28 joists to lie in the bottom, under the planks, every joint being
9 inches square and 25 feet long so every joist will contain 16 feet and one
quarter of a foot which will amount to 9 loads of square timber.

Item, for the bottom 24 planks, every plank being 70 feet long, and 1 foot
broad and 3 inches thick, so every plank will contain 17+ feet, which will
amount to 8 loads and 20 feet of square timber.

Item, there must be 2 ground-sills, every ground-sill of 70-foot length and
1 foot square, which will contain 140 feet of square timber, that is 2 loads
and 40 feet.

Item, for every side there must be 28 studs, every stud being 12 inches
broad and 8 inches thick and 10 feet high, so every stud will contain 7 feet
in square timber, which will amount to 8 loads.

Item, there must be 2 pieces of timber to lay upon the said studs, every piece of 70-foot length, and 1 foot square, which is 2 loads and 40 feet.

Item, every side will require 10 planks, every plank being 70 feet long, 1 foot broad and 3 inches thick, so every plank will contain 17+ feet, which will amount to, for both sides, 7 loads of timber.

Item, the 4 gates with the posts to hang them upon, and certain 'Lande Keyes'† for the same, will require 6 loads of timber.

$$\left.\begin{array}{l}\text{Total loads of timber} = 44 \\ \text{every load rated at 10s. per load}\end{array}\right\} \text{£}22$$

Item, the sawing work of all the aforesaid timber will amount to 56 hundred at 20d. the hundred £4 18s. 4d.‡

Item, the carpenters work for the same lock will cost 20 marks§ so that the carpenter be put to no other charge but only the framing, and setting up of the same.

Item, if the sides and ends of the said lock be made of stone or brick, then the timber that shall go to the same will amount to 24 loads or thereabouts.

Item, the workmanship of the same wall, being 10 feet high, and 3 feet thick will cost 16s. a pole, so that the mason be put to no other charge, but only the raising of the same wall, and the same work will contain by estimation 10 poles, every pole being 16 feet.

Item, there will go to the same wall 8 loads of lime which will cost 16s. 8d. the load, which is £6 13s. 4d.

Item, there must be to the same work 16 loads of sand at 2s. the load, 32s.

Item, the heightening of the high bridge at Waltham to raise it 2 feet higher will cost by estimation £4.

Item, the charges of the cutting of the new cut from the mill river to the river of Lee, which contains 36 poles, every pole rated at 26s. 8d. the pole, which will amount to £48.

		£	s.	d.
if it be done	Timber, 44 loads at 10s.	22	0	0
with timber	Sawing & carpenters' work	18	5	0
		40	5	0

the same	Timber 24 loads	12	0	0
with stone	Lime 8 loads at 16s. 8d.	6	13	4
	Sand 16 loads at 2s.		32	0
	The walling	8	0	0
	Carpenters' work			
	Stone			
		28	0	0 [*sic*]
	The heightening of the bridge	4	0	0
	The charge of the cut	48	0	0
		52	0	0

Explanation of symbols (provided by Dr Norman Smith)
* One 'lode' of timber equalled 50 cubic feet.
† 'Lande keyes' probably means timber-work anchored into the ground to brace the posts; or, conceivably, it refers to some wooden device to open and close (hence key) the gates.
‡ The amount given is in fact incorrect, it should be £4 13s. 4d.
§ One mark was equal to two-thirds of a pound sterling, i.e. 13s. 4d.

Notes

1. 13 Eliz. 1 c. 18.
2. Bodleian Library, Rawlinson MSS, Essex 11.
3. John Norden's 'Description of Middlesex', British Library, Harleian MS 570.
4. For a full account of events along the Lee in the Elizabethan period, see my account 'The River Lee: A Tudor Experiment in River Navigation' shortly to be deposited with the Stratford Reference Library.
5. See A.W. Skempton, 'Canals and River Navigation before 1750' in Vol. 3, *A History of Technology* (ed. C. Singer *et al.*, London 1957, p. 456); Philip Chilwell De la Garde, 'Memoir of the Canal of Exeter 1563–1724' in the *Proceedings of the Institution of Civil Engineers*, Vol. IV, 1845, pp. 90–102; George Oliver, *The History of the City of Exeter*, Exeter 1861–84.
6. William Vallans, 'A Tale of Two Swannes. Wherein is comprehended the original and increase of the River Lee' [1590]; in J. Leland, *The Itinerary*, Vol. V, Oxford 1710.
7. In 1594–5 a manuscript book entitled 'Proceedings in the Star Chamber' was compiled by William Harte. It is now kept in the Guildhall Record Office. Besides giving the 1594 Star Chamber case in great detail, it also provides copies of many other ancient documents relating to the Lee. Folios 174–7 provide a record of a Commission of Sewers' meeting held in the Star Chamber on 19 November 1577. This is the only remaining evidence to the Commissioners' decisions relating to Waltham.
8. *ibid*
9. *ibid*
10. British Library, Landsdowne MS 22 no. 48. Thomas Fanshawe wrote to Lord Burghley that he had just visited Waltham 'to see whether passage might not be made from the mylne Dame thorough the towne by some of the channelles there'.
11. Public Record Office, State Papers Elizabeth, Domestic Series, Vol. 109 no. 33.
12. *The Oxford English Dictionary* mentions a statute in 35 Eliz. 1 which defined a pole as being 16½ feet. If this measurement can be applied to this particular estimate then the length of the new cut would have been 176 yards.

13. Harte, *op. cit.*

14. I would like to thank Dr Norman Smith for his helpful discussion and Dr Denis Smith for drawing figures 1 and 2.

15. British Library, Landsdowne MS 32 no. 41.

16. Public Record Office, State Papers Domestic: Addenda 1580–1625, Vol. 27 no. 6.

17. British Library, Landsdowne MS 32 no. 40, paginated 109.

18. British Library, Landsdowne MS 32 nos. 35 and 41.

19. Harte, *op. cit.* folios 1–38 deal with the Star Chamber case.

20. British Library, Landsdowne MS 77 no. 16; British Library, Add MSS 33576 fo. 63; Public Record Office, Exchequer 178/4965; and Northamptonshire Record Office, WC 244.

21. Northamptonshire Record Office, YZ 6097.

22. Northamptonshire Record Office, WC 244.

23. P.J. Huggins, 'Excavations of a Medieval Bridge at Waltham Abbey, Essex in 1968', *Mediaeval Archaeology*, XIV 1970, pp. 126–47.

24. Public Record Office, State Papers Elizabeth, Domestic Series, Vol. 109 no. 33.

Parkesine and Celluloid:
The Failure and Success of
the First Modern Plastic

ROBERT FRIEDEL

Historians have paid remarkably little attention to the development of the novel materials that began to appear in science and industry in the second half of the nineteenth century. These new substances, however, are as characteristic of modern technology as the more spectacular developments in transportation, communications, manufacturing and power that are the classic subjects of historical attention. They also represent as marked a departure from the past as any other technological phenomenon of modern times. This discovery or development of new metals, plastics, fibres or the like has expanded the material capabilities of industry to a degree comparable to the expansion of the forms of energy represented by electricity or internal combustion. And yet, for all the attention paid to the technological 'revolutions' of the past two centuries, scholarship devoted to the beginnings of these new materials and their expanding exploitation is quite scanty.

A good example of this neglect is the history of plastics. The first artificial plastics made their appearance shortly after the middle of the nineteenth century, and from that time the swelling stream of comparable artificial substances has been one of the most conspicuous products of chemical industry. Nevertheless, a comprehensive history of these materials has yet to appear, and general histories of technology, or even of chemical technology specifically, give only cursory attention to the plastics industry.[1] This neglect is unfortunate, not only because it leaves unchronicled a large and highly visible industry, but also because the development and exploitation of novel materials raise interesting and significant questions about how new technologies are conceived and made economically viable. New machines, for example, are often obvious improvements in older ways of doing things. New materials, on the other hand, present novel combinations of properties whose functions may be purely conjectural. Why such combinations of properties should have been created at all is perhaps the most immediate question raised by the invention of a new substance. But equally insistent is the question of how a novel substance is integrated into the technological matrix around it. Such was the novelty of materials like the first plastics that there was considerable difficulty in determining their uses. Inventors and entrepreneurs not only had to solve technical problems in the manufacture of their products, but also had to develop significant applications, establish viable markets and, finally, shape the economic and social status of their invention. This is a pattern of development which is hardly unique to new materials, but the problems are presented particularly starkly in the case of the first plastics.

Celluloid, the first artificial plastic to achieve any measure of technical or commercial importance, was introduced about 1870. That introduction was preceded by almost two decades of attempts to create and market such a material. For millenia men had been using materials that were in some sense 'plastic'. Metals, clays and glasses — any substance that is sometimes shaped by moulding — are in some sense 'plastic', but they are never called 'plastic'. In the first half of the nineteenth century, however, there began to appear other materials, much closer in their properties and in the manner of their working to the modern plastics. The most important of these 'natural plastics' were india rubber and gutta percha — both made from the gums of trees found far from the industrial centres of Europe and America. The significance of celluloid lay in its independence from these exotic sources. Made from materials completely devoid of plasticity in their natural state, celluloid was truly an artificial substance, totally the product of chemical manipulation and well-defined processes of handling and finishing.

The story of the invention of celluloid begins with the discovery in 1846 of nitrocellulose by Christian Friedrich Schönbein in Basel. A number of French investigators had produced similarly nitrated organic products earlier, but Schönbein discovered a method for producing nitrated cellulose in quantity and also recognized some of the significant properties of the substance, particularly its explosiveness and its solubility. It was the potential of nitrocellulose as an explosive that immediately caught the world's attention, though it was years before the material was of military importance. It did not take long, however, for curious dabblers to explore the possibilities presented by the fact that cellulose nitrated somewhat less than 'guncotton' could be dissolved in such common solvents as alcohol and ether. When so dissolved, nitrocellulose produces a clear, syrupy fluid that, poured out and allowed to dry, results in a thin transparent film. This film was the progenitor of celluloid.

The first application of collodion, as the solution of nitrocellulose in a mixture of alcohol and ether was called, was announced less than a year after Schönbein introduced nitrocellulose. In January 1847, a Boston medical student named J. Parker Maynard suggested that collodion could be used as a 'sticking plaster' to provide a convenient water proof covering for wounds.[2] The popular identification of collodion, however, was not with surgery — though this lasted well into the twentieth century — but with photography. In 1851, F. Scott Archer introduced his 'collodion process' or 'Archerotype'. While this was not the first suggestion of collodion's use as a vehicle for photosensitive materials, Archer's wet plate process was the first practical alternative to the daguerreotype and albumin processes and was for more than twenty years the dominant photographic mode.[3]

The step from the syrupy collodion and the thin film it created to a solid, mouldable plastic was not made easily or quickly. The first progress in making that step was the work of Alexander Parkes (1813–90), metallurgist, chemist and inventor extraordinary of Birmingham. Parkes was trained and employed by manufacturers in the non-ferrous metal-working industry of Birmingham, and from them acquired a wide-ranging knowledge of different materials. Parkes made a name for himself working with precious metals, india rubber and gutta percha, as well as with nitrocellulose.[4] In the early 1850s Parkes began experimenting with collodion, and in 1855 received his first patent (British Patent 2359 of 1855) for using the material. Therein he announced his aim 'to employ collodion or its compounds for manufacturing purposes generally'. The patent then described the use of collodion or collodion-based compounds for water-proofing fabrics, leather, plaster or wood; for decorating fabrics or paper; for sheets that would be useful for bookbinding, button-manufacture, 'and other applications where a hard, strong, brilliant material is required'; and for moulding ornamentations. 'A hard, strong, brilliant material' that could be cut or moulded — there could be no more explicit description of a plastic.

Despite his patent of 1855, Alexander Parkes did not attempt to introduce his invention until 1862. At the Great International Exhibition held in London that year, Parkes displayed a modest collection of medallions, buttons, combs, boxes and similar small items. Classed in the Exhibition among the 'Vegetable Substances Used in Manufactures, &c.', the display was described in the *Catalogue* as simply 'Patent Parkesine of various colours; hard, elastic, transparent, opaque, and waterproof'. Parkes was awarded a medal for 'quality', but his efforts evoked no further comment from the Exhibition's official sources.[5] Parkes could not have been too disappointed, however, for he did catch the eye of George Spill & Company, manufacturers of water-proof cloth in East London. George Spill and his brother Daniel approached Parkes about the possibility of using the water-proofing qualities of his invention. Rather than simply licensing the Spills, however, Parkes persuaded them to join with him in an effort to develop the full range of commercial possibilities offered by 'parkesine'. In 1864, Parkes began work at the Spills' factory with the aim of developing means to produce parkesine and fabricate products from it on a commercial scale.[6]

Parkesine was a commercial failure. Despite the establishment of a well-backed company and the operation of a sizeable factory, the material did not sell. This failure had several sources — both technical and economic. At root was the fact that Parkes never developed a dependable formula for a nitrocellulose plastic. The patent of 1855 described solutions of guncotton in 'vegetable naphtha, alcohol, methylated or other spirits, or other solvents'. To this was to be added gums, resins, stearine, rubber, gutta percha, or colouring agents. Nine years later, on the eve of beginning manufacture, Parkes received another patent for his material (British Patent 2675 of 1864) which showed that he was still uncertain of the proper formula for a useful substance. Therein he described a variety of solvents for nitrocellulose, and continued to prescribe the addition of oils, resins or gums. The key to the production of a successful nitrocellulose plastic was the use of a non-volatile solvent that also acted as a plasticizer. This crucial element turned out to be camphor — one of the dozens of substances Parkes dabbled with in his experiments. Parkes was not able to recognize the importance of camphor because of his persistence in using liquid solutions of nitrocellulose. Parkes could never succeed without a major shift in his perception of his material.

The technical problem was not the only one, however. The parkesine shown at the 1862 Exhibition had, after all been praised for 'quality'. Why then did the output of the Parkesine Company become notorious for warping and shrinking, for being too unreliable for even the cheapest sort of goods? The failure of parkesine was a failure not only of chemical technology, but also of commercial vision. Rather than attempting to perfect appropriate products that could be made from carefully prepared parkesine, all efforts were directed toward making the material more cheaply. In a lecture to the Royal Society of Arts in London in 1865, Parkes declared that he could produce 'any quantity' of parkesine for less than one shilling per pound. This economy (requiring 'no less than twelve years' labour and an expenditure of many thousand pounds' to achieve) was effected by cheapening the materials of manufacture.[7] Parkes devised means of producing nitrocellulose from cotton, linen and paper waste. He used special devices to recover the solvents of the nitrated cellulose and adopted cheaper solvents. Simply to move parkesine from the laboratory into the factory did not require these sorts of efforts. This emphasis on cheapening the material indicated that parkesine was not being viewed as a material to extend the range of the artist or the craftsman, as Parkes had once pictured it, but was rather thought of as an all-purpose industrial commodity.

The efforts that began in 1866 to recruit capital for the newly incorporated Parkesine Company confirmed this shift to a rather mundane perception of the material. The *Prospectus* issued for the new company described parkesine's potential in especially proasic terms:

> Parkesine remains unaffected by the heat of tropical climates, and is not subject to become brittle by cold, nor impaired by oxidation.
>
> It may be made opaque or transparent, hard or flexible, and of any

colour. In its hard form, it can be applied to carding, roving and spinning rollers, or bosses for cotton, linen, silk and woollen machinery, insulators for telegraph poles, telegraphic and philosophical instruments, book binding, picture frames, panelling for carriages, and works of art.

In its flexible condition it may be extensively used for insulating telegraphic wire, for the manufacture of tubing, for coating textile fabric, and fuses for blasting purposes, &c.; and, in its fluid state, as varnishes and paint for general purposes, and more especially for coating iron ships.

The practicability of shallow and deep sea telegraphy having been conclusively and successfully demonstrated, the demand for insulating materials has of late years greatly increased, and must now necessarily be very great; and, the supply of ordinary insulating materials becoming precarious, the Company is justified in anticipating that an extensive and profitable trade will arise from the application of Parkesine to this purpose.[8]

In 1866 both Britain and America were ecstatic over the first permanently successful Atlantic cable. Little wonder that a small group of entrepreneurs with an unknown product should wish to link Cyrus Field's success to their own prospects. Unfortunately, parkesine was in fact a very poor material for telegraph cables. Not only did parkesine shrink too much to be useful, but it was never, even in the form of celluloid, cheaper than gutta percha.

More important than fruitless speculation about tapping the cable market was the new tone of the 1866 *Prospectus*. With the exception of a quick reference to 'works of art', the idea of parkesine as a beautiful material for making fine things was absent. In 1862 Parkes had been at pains to show his invention as a new vehicle for the artist. The appeal to financiers and industrialists four years later, however, was in terms of the common and practical — spinning rollers and fuse coatings. Parkes was not unaware of this significant shift in perspective. Years later he testified:

> ... when I was manufacturing Parkesine myself my principal object was to make works of art and things of high class value but when I negotiated this thing on a large scale everybody connected with the movement from the commencement of Messieurs Spill and Company urged the necessity of making Parkesine as cheaply as possible — it was to be dealt in in tons instead of pounds.[9]

Despite his implied protest, Parkes himself did not resist this change. His presentation before the Royal Society of Arts, after all, included enthusiastic references to the cheapening of parkesine.

Simple cheapening of the material was not the only unfortunate aspect of the effort to commercialize parkesine. An equally serious problem was the lack of a clear vision of the marketable uses of the substance. Not only were impracticable uses put forward, but too broad a range of possibilities was confronted. The great jumble of suggestions contained in the 1866 *Prospectus* was one symptom of this problem. Another appeared in the pamphlet Parkes prepared to accompany his company's display at the Paris Universal Exhibition in 1867. Therein Parkes offered this explanation for his invention:

> For years previous to 1850 I was strongly impressed with the importance and want of a substance to take the place of certain natural productions, such as ivory, tortoiseshell, india rubber, gutta percha, &c., &c., and in the year 1850 I directed my attention to the peculiar properties of Pyroxyline and similar substances.[10]

While this may have been an honest statement, the confusion of luxurious materials like ivory or tortoiseshell with vulgar substances such as rubber (not to mention the '&c., &c.') could only promote uncertainty about the useful functions of parkesine. The Paris pamphlet went on to boast that 'from its great beauty, unchangeableness, and variety, the substance is capable of being used for a greater number of purposes than any other material known at the present time'. The objects on display included cameos, book covers, combs, brushes, tubing, knife handles, buttons, purses, and hopeful, but not much touted, examples of coated telegraph wire.

Parkes and his partners in the Parkesine Company generally attributed their failure — the assets were liquidated in 1868 — to technical difficulties with their product. There can be little question but that Parkes' formulas and techniques were not adequate to the factory production of a nitrocellulose plastic. Nevertheless, the failure to understand that a new material's functions and outlets required just as much invention as the substance itself was as fatal to parkesine as technical deficiencies. The using of a new substance was as great a challenge to the imagination as creating it.

The lessons to be learned from parkestine's failure become even more emphatic in the light of celluloid's success. Not only did the invention of celluloid solve the technical problems that always plagued Alexander Parkes, but its commercialization illustrated clearly the sorts of approaches that were necessary to introduce a new material into an economic and technical environment quite unused to novel substances. The commercial success of celluloid was not an easy or straightforward achievement. A great deal of faith and persistence was necessary before the material could be considered economically viable. The conservative approach of the makers of celluloid stands in marked contrast to the dramatic but abortive claims made for parkesine, and is perhaps the most illuminating aspect of celluloid's commercial survival.

Celluloid was essentially a mixture of nitrocellulose and camphor. A mechanical mixture, aided by the addition of small amounts of volatile solvents (such as alcohol) to speed up the mixing, celluloid differed from parkesine in that it never went through a liquid stage in its manufacture. Moderately nitrated cellulose (generally from paper) was finely shredded and mixed with well-divided camphor for many hours — with just enough liquid solvent to soften the mixture. The resulting mass was pressed, broken up, pressed again and formed under rollers or in blocks. While the large-scale manufacture of the substance required numerous specialized machines and processes, the entire operation was a relatively simple one — involving only a small number of materials and little chemical sophistication in handling them.[11]

The inventor of celluloid was John Wesley Hyatt (1837–1920), a printer and general tinkerer of Albany, New York. In 1863, so the oft-repeated story goes, Hyatt's attention was caught by the offer of a prize of $10,000 from the firm of Phelan and Collender in New York City for the patent rights to an ivory substitute suitable for the manufacture of billiard balls. Attacking the problem in a conventional manner, Hyatt proceeded to make up various well-known plastic compositions, such as a combination of pressed wood pulp and gum shellac that was then popular for daguerreotype cases. These materials did not produce a satisfactory billiard ball, but Hyatt did set up a company to use the compound for the manufacture of small articles like checkers, dominoes and so forth.

An important result of Hyatt's work with these early plastic compositions was the acquiring of a familiarity with processes for moulding plastic compounds under heat and pressure. Other authors have not generally pointed out that Parkes and the Spills lacked such familiarity when they began their work with collodion. These earlier workers thought more often in terms of making liquid collodion solid, rather than the solid nitrocellulose mouldable. This was the most critical difference in the methods of Hyatt, whose own description of how he came to produce a plastic from nitrocellulose is the best one:

> From my earliest experiments in nitrocellulose, incited by accidentally finding a dried bit of collodion the size and thickness of my thumb nail, and by my very earnest efforts to find a substitute for ivory billiard balls, it was apparent that a semi-liquid solution of nitrocellulose, three-fourths of the bulk of which was a volatile liquid and the final solid from which was less than one-fourth the mass of the original mixture, was far from being adapted to the manufacture of solid articles, and that I must initially produce a solid solution by mechanical means.[12]

In 1869 Hyatt took out several patents. Two of these related to the use of collodion. The first, U.S. Patent 88634, was for applying coats of collodion to composition billiard balls, but the second was the earliest indication that Hyatt had seriously turned his attention to the problem of producing a plastic from nitrocellulose. U.S. Patent 91341, for an 'Improved method of making solid collodion', was significant for the emphasis that it placed on the use of high pressure. The product described by Hyatt was not celluloid, but a composition that used solid fillers such as ivory-dust, asbestos and the like. Hyatt continued to pursue the goal of making a truly solid collodion, and, seeing references to the usefulness of camphor in reducing the shrinkage of collodion, he struck upon the idea of attempting to mix nitrocellulose and camphor together directly, with a negligible amount of volatile solvent present. He described the successful result in U.S. Patent 105338 (issued 12 July 1870) as 'a solid about the consistency of sole-leather, but which subsequently becomes as hard as horn or bone by the evaporation of the camphor'. This was celluloid, though the name was yet to come.

The material invented by Hyatt did not change much in subsequent years. It was tough, uniform and resilient. It possessed a high tensile strength and was resistant to water, oils and dilute acids. Celluloid was fairly light, having a specific gravity of about 1.4. It took a high lustre and colours and colouring effects well. It could be easily worked, and at room temperature could be sawed, drilled, planed, turned, buffed and polished. Celluloid could be moulded at a temperature just below that of boiling water. At higher heats (above 365°F), however, it would start to decompose, and it was, of course, highly flammable. Still, the combination of attractive properties in terms of working, appearance and durability, and the cheapness and ready availability of its constituent materials, made the material an irresistible challenge to the imagination and enterprise of the inventor–entrepreneur from Albany.

The Albany Billiard Ball Company was one of John Wesley Hyatt's first commercial ventures. Contrary to most accounts of Hyatt's work, the products of that company were almost certainly not made of celluloid. The new material simply did not have the proper combination of density and hardness to substitute for ivory in billiard balls. Hyatt's company manufactured balls with a shellac or shellac-composition core sometimes coated with collodion or thin celluloid. Even this minor use of celluloid was not particularly successful, as indicated by the story Hyatt once told of a Colorado billiard saloon-keeper who wrote to complain that occasionally a sharp collision of collodion-covered balls would produce a report like that of a percussion cap, causing every man in the establishment to draw his gun.[13] While the story of celluloid's origin as the product of the search for a cheaper billiard ball is in some sense true, it is important to note that celluloid did not and could not succeed as a billiard ball material. While celluloid was eventually much used as an imitation ivory, in the case of billiard balls the differences between celluloid and ivory were more important than the similarities. In such a use the qualities of the original material rigidly defined the necessary properties of a substitute. Versatile as celluloid might be, it could be manufactured only within narrowly defined ranges of density and elasticity. Only in the twentieth century, with the development of phenolic plastics that could be made in a wide range of compositions, did there appear a substitute for the ivory billiard ball.

The first enterprise established for the production of celluloid products was the Albany Dental Plate Company, begun by John Wesley and his brother Isaiah Smith Hyatt in 1870. For almost two years the Hyatt brothers devoted themselves to making celluloid a successful substitute for hard rubber in dental plate manufacture. Once again the approach of the Hyatts stood in marked contrast to that of Parkes and his English colleagues. The commercialization of celluloid was undertaken on narrow and carefully chosen ground. The opportunities perceived in the dental plate industry derived not from the scarcity of the material predominantly used for plates — hard rubber — nor from inherent difficulties met by dentists using rubber, but from monopoly. The holders of the Goodyear Company's patent on the use of rubber for dental plates charged a royalty to every dispenser of hard rubber plates, a charge apparently found irksome, if not onerous, by dentists everywhere.[14] The experience of the Hyatts in this first extended effort at the technical and economic exploitation of celluloid was a further reflection of the problems of introducing a novel material into nineteenth century markets.

During the first half of 1871 the Albany Dental Plate Company put the following notice in several numbers of the *Dental Cosmos*, the journal of a dental supply house:

> We take great pleasure in announcing to the Dental Profession, that we are in possession of a newly-invented and patented material for Dental Plates or bases for artificial teeth, that cannot fail to delight every dentist who desires a better material for the purpose than hard rubber.
>
> This base consists of a new and peculiar composition of solid collodion, which possesses all the advantages that have ever been hoped from collodion or pyroxyline, while it is entirely free from the difficulties heretofore experienced in manipulating that substance, as well as from liability to shrink and change form after being made into artificial plates.[15]

This was the first announcement of a celluloid product. The advertisement went on to list the advantages of the new plates: lighter and stronger than hard rubber, truer in colour, free from unpleasant taste, acid resistant, harmless in the mouth (lacking the mercury used in colouring rubber plates), more easily and quickly fitted, and more comfortable to wear than rubber plates. The blank plates were offered at $1.00 apiece and the complete apparatus for moulding the plates was priced at $7.00. Price was not one of the touted advantages of celluloid. Contemporary advertisements for rubber gave the price of dental rubber at about $2.50 per pound. A typical hard rubber plate weighed no more than half an ounce and thus contained no more than $.08 worth of rubber.[16] Those irritating royalties would have to be added to the cost of using rubber, but even so, it is clear that celluloid could only be a luxury alternative.

Unfortunately, for the Hyatts, technical difficulties made their product less than satisfactory. A little while after the appearance of the advertisement quoted above, there appeared another item in the *Dental Cosmos* from the Albany Dental Plate Company, this one reflecting considerable concern over the fate of their product. In the form of a letter from Isaiah Smith Hyatt, the piece began:

> I have received numerous letters from members of the Dental profession who are using the 'Celluloid Base', and who, while speaking of their successes, also give instances of failure or of imperfections existing in some of the plates which they have used . . .
> The difficulties referred to may be summarized about as follows:
> 1. Some of the plates have had a strong camphoric or pungent gummy taste.
> 2. Some have become soft in the mouth, or sufficiently so for the teeth to loosen.
> 3. Plates have warped after having been adjusted in the patients' mouths.
> 4. Plates have been found that were flaky or laminar.
> 'What is the matter?' dentists ask. 'We had great hopes of the new base, but if it works this way, we shall, in great disappointment, be obliged to go back to rubber'.[17]

Hyatt went on to explain that these difficulties were met with only by a few dentists and that they were the unavoidable symptoms of a new manufacture. They were said to have been corrected by variations in the composition of the plates and by greater care in their manufacture. The advertisement was, it should be added, probably the first using the name 'celluloid', and all subsequent notices by the Hyatts used the new name freely.

Despite the fact that celluloid dental plates continued to be made at least until after the end of the century, they never replaced hard rubber as the most important material for plates. The problems acknowledged by Hyatt were never successfully overcome. One French author spoke of the abandonment of celluloid in favour of hard rubber 'on account of the pronounced taste of camphor in the dental plates, of the apprehension of practitioners over the inflammability of the product and of its warping under the influence of heat'.[18] Another reason that celluloid was not able to compete successfully with hard rubber was that its price always made it the more expensive substance, especially after the end of Goodyear patent control over dental rubber. Its one major advantage was its colour. Hard rubber was naturally opaque brown and making plates from it that had any kind of acceptable gum-like colour required the use of large amounts of agents like zinc white and vermilion, this last being a mercury compound whose presence in the mouth was often a cause of concern.[19] Celluloid, on the other hand, could be coloured the proper shade of red with much smaller amounts of vermilion or even with completely safe dyes, and was hence considered a healthier material. Simply being prettier and healthier, however, did not make celluloid a truly significant replacement for hard rubber in dentistry.

Despite some difficulties with dental plates, by 1871 the Hyatts felt that their experience warranted expansion of their operations. In January of that year the Celluloid Manufacturing Company was organized in Albany with an initial capitalization of $60,000. The stated purpose of the company, which absorbed the dental plate operations, was to manufacture and sell celluloid in a semi-finished form — in rods, sheets, tubes,etc. The formation of the new company also marked the beginning of intensive efforts to recruit the financial backing that would be necessary if the manufacture of celluloid was to become more than a small, localized affair. These efforts bore fruit when late in 1872 the Celluloid Manufacturing Company moved from Albany to a factory on Ferry Street in Newark, New Jersey. A well-known New York City business-man, Marshall Lefferts, became the president of the company, and the promoters of celluloid embarked upon a careful and systematic campaign to create a dependable and significant market for their product.[20]

This campaign was as much responsible for the eventual success of celluloid as Hyatt's technical breakthroughs. Nothing distinguished the American experience with celluloid more from the English efforts that went before than the gradual and systematic introduction of the material into carefully chosen markets. While the Hyatts and their colleagues were as convinced as had been Parkes and the Spills that the new plastic had wide-ranging possibilities, they were extraordinarily conservative in the introduction of new celluloid products. Where Parkes had touted widely the versatility of parkesine, picturing the material as an inexpensive substitute for a host of natural substances, the makers of celluloid began by cautiously testing one market after another, exploiting successes readily, but taking care not to be tainted by failures. The story of this effort is complex, but even in its outlines it reveals a great deal about how a novel material could be integrated into the technological and economic life of the late nineteenth century.

The mixed experience with dental plates persuaded the celluloid makers to devote the Newark factory to the production of unfabricated celluloid, primarily but not exclusively in the form of sheets. The making of consumer goods from celluloid would be the business of licensees, some of whom would be established businesses that had been persuaded to use celluloid in the manufacture of their products while others would be firms set up with the encouragement of the Celluloid Manufacturing Company. The list of licences granted for the production of celluloid goods during the 1870s reflects the pattern of commercialization that succeeded in establishing celluloid as a widely-accepted material by the end of the decade. The following list is derived from testimony given under oath in 1880 in the course of litigation between Daniell Spill and the Celluloid Manufacturing Company over Spill's claim of patent rights to the Hyatt's product.[21]

Date of issue	Licensee
1 Oct. 1872	Samuel S. White (dental supplies)
12 Feb. 1873	Celluloid Harness Trimming Company
26 Jan. 1874	Edward C. Penfield (medical supplies)
20 Nov. 1874	Meriden Cutlery Company
21 Nov. 1874	I. Smith Hyatt (later, Celluloid Brush Company)
1 Sept. 1875	Emery Wheel Company
18 Sept. 1875	Spencer Optical Manufacturing Company
22 Dec. 1875	Celluloid Novelty Company
1 Mar. 1877	Albany Billiard Ball Company
9 Mar. 1878	Celluloid Waterproof Cuff and Collar Company
23 Oct. 1878	Celluloid Hat & Trimming Company
31 Oct. 1878	Celluloid Fancy Goods Company
30 Nov. 1878	Celluloid Shoe Protector Company
12 Dec. 1878	Celluloid Piano Key Company
1 Apr. 1879	Celluloid Veneer Company
1 Mar. 1880	Celluloid Surgical Instrument Company

The gradualism practiced by the Newark Manufacturers is apparent from this list. It seems natural that the first licence should be to the S.S. White Company, the Philadelphia dental supply firm that had handled the output of the Albany Dental Plate Company. The Harness Trimming Company was the first of numerous small firms established in Newark to work celluloid into marketable forms. It was first set up at the same address as the Celluloid Manufacturing Company and made rings, hooks, buckles, rosettes, and other harness trimmings, competing largely with the much more expensive ivory or the somewhat unsatisfactory (in appearance if not in performance) hard rubber.[22] Such a product represented a humble but effective way of challenging more traditional materials.

Penfield & Company was a Philadelphia medical supply house which may have been influenced by the S.S. White Company in its decision to try out celluloid for the manufacture of truss pads and coatings for truss springs. Here too celluloid was competing with hard rubber, and it would have to be considered a luxury item, preferable for its better flesh colouring and its flexibility in cold weather. Knife handles and brushes and combs had been among the few important products of the parkesine manufacturers, and their early appearance in celluloid was to be expected. In cutlery, celluloid offered another material in a field that already had a great number. The ease with which celluloid could be coloured and worked probably made it a very attractive replacement for such natural materials as bone, horn, shell and ivory. While it is not possible to say precisely to what extent or how fast celluloid gained acceptance as a material for knife handles, as late as 1938 it was reported that cutlery manufacturers in Britain were 'emphatic' in maintaining the importance of celluloid to their trade.[23]

While it was the Celluloid Manufacturing Company's policy to leave the making of final products to other firms, the celluloid producers themselves were still instrumental in developing the techniques for exploiting celluloid's properties. Between 1869 and 1891 there were issued to John Wesley Hyatt sixty-one patents relating to the manufacturing and working of celluloid. While many of Hyatt's patents involved formulae and machines for celluloid production and finishing, a number were for the making of specific articles, among them collars, cuffs and combs. Other important individuals in the Celluloid Manufacturing Company took out patents during this period, many of these also reflecting a concern for developing celluloid products. Marshall C. Lefferts, son of the Marshall Lefferts who directed the company during its first years in Newark and later president of the company itself, was granted patents for such products as syringes, spoons and forks, dolls, stays and ice pitchers. As might be expected, many of the processes developed by the Celluloid Manufacturing Company and by other celluloid producers were of little or no commercial importance, but the very fact of their patenting indicated that product development was always important to the makers of the material.

All but two of the firms licensed by the Celluloid Manufacturing Company to fabricate celluloid were specialized — in the business of making and selling a very narrow range of consumer goods. Specialization of product meant specialization of labour — the ability to recruit and employ workers skilled in producing a traditional product like combs or piano keys, and specialization of marketing — the opportunity to compete with traditional materials on carefully selected ground. Instead of introducing celluloid to a multiplicity of markets as a suitable replacement for a wide range of different materials, celluloid was presented as a good replacement for starched linen in collars and cuffs at one time, and a substitute for ivory in piano keys at another. The usefulness of the material as a substitute could be carefully judged by both consumer and producer as each product was introduced. Sometimes the effort would be a failure: the Hat & Trimming Company was never able to sell its product, nor was the Shoe Protector Company. In other cases the step-by-step approach produced surprising successes, as in the case of the Waterproof Cuff and Collar Company, which was able to make a novelty into an acceptable article of dress. It came at a time when the number of poorly-paid clerical workers was greatly expanding and provided welcome relief to a class that could not readily afford a daily change of linen — celluloid was easily sponged. Its use as a substitute for fabric continued as long as the use of removable collars and cuffs.

The two licensed celluloid fabricators who did not have narrowly defined products — the Novelty Company and the Fancy Goods Company — were perhaps the most characteristic celluloid fabricators of all. The output of these firms was important in shaping both the image and the market of celluloid. The best account of the articles produced by these two companies comes from two ledgers kept by the Celluloid Manufacturing Company, titled 'Licenses, Contracts and Patents', listing various agreements made by the company in the years from 1872 to 1878. From just the index to these volumes it is possible to construct a summary of the products that the Novelty Company and the Fancy Goods Company were licensed to produce:[24]

The Celluloid Novelty Company

Armlets	Watch chains	Shawl pins
Breast pins	Ear-rings	Sleeve buttons
Bracelets	Jewelry	Shirt studs
Crosses	Necklaces	Scarf rings and pins
Charms	Pendants	

The Celluloid Fancy Goods Company

Checkers	Bows and scarves	Thimbles
Cribbage boards	Card and jewelry receivers	Tape measures
Dice boxes	Salt cellars	Thermometers
Key rings	Soap dishes	(fancy)

These lists are probably not exhaustive — such popular celluloid items as razor handles or shoe horns do not appear but might very well have been produced by these companies. Still, the lists are sufficient to convey a general image of the variety of small articles being made out of celluloid by the end of the 1870s.

Such diversity brings to mind the products of the parkesine makers, but the Newark companies did not have the problems of the English manufacturers. Even here, a different marketing strategy helped the companies to avoid the fate of Parkes. All of the articles sold by the Novelty Company were items of personal adornment, and thus could be marketed together through fairly narrow channels. Cheap jewelry was not a new idea in the late nineteenth century, and celluloid items, despite their diverse forms, could be presented as new sorts of inexpensive adornment, superior in appearance to hard rubber or other cheap substance, and considerably cheaper than coral, amber or pearl. Similarly, despite the diversity of articles made by the Fancy Goods Company, they fit easily into a well-defined segment of the market — fancy 'notions'. These were the small, eye-catching items that many clothing or houseware or stationery retailers would put out to attract their customers' pocket money. In these products celluloid might be replacing not only semi-precious substances, but also wood, glass or metal. The Fancy Goods Company licence was not issued until late 1878, almost six years after celluloid production began in Newark. By this time celluloid had possibly received enough exposure to be used in forms not commonly met with in the materials which it usually imitated, such as ivory or tortoiseshell. It certainly should not be inferred, however, that celluloid was not still an essentially imitative material. The breakaway from imitation could be achieved only by the creation of a new product made possible by the new material. This achievement, embodied by photographic film, lay well into the future.

In the meantime considerable effort was expended in producing celluloid imitating a wide variety of natural substances. The first and most important imitation was of ivory. Whatever the difficulty may have been in making a satisfactory celluloid billiard ball, for other uses of ivory the material was eminently suitable. A successful ivory imitation had to duplicate not only the colour of the natural substance, but also its peculiar and distinctive graining. With celluloid this could be done by putting sheets of varying shades of white and of varying thicknesses together, bending and folding the stacked sheets and repeatedly putting them through heated rollers. A look at a collection of old celluloid products, such as that in the Smithsonian Institution's National Museum of History and Technology, shows an overwhelming preponderance of imitation ivory. This extends even to many articles that would never have been made of natural ivory, such as visiting cards and vases.

The disposition toward the ivory form was both because of and despite the versatility of celluloid. A material like hard rubber was difficult to make in attractive colours due to its natural darkness. Hence black became the colour associated with rubber products, and even jewelry was made in shiny black 'vulcanite'. Rubber was forced to create its own image. On the other hand, celluloid was under no such restraints; it could be very easily made to imitate ivory or coral or amber or any of a wide variety of natural substances. It could also be coloured unlike other materials — in any colour of the rainbow, in a wide variety of transparent and opaque forms. Freed from the constraints that hampered a material like hard rubber, celluloid makers found it easier to popularize their products in forms both familiar and admired, rather than exploit the creative possibilities of their material. This was true not only in the first years of celluloid manufacture, but persisted for decades, even when rival plastics began to appear.

While ivory was certainly the most popular imitative form of celluloid, it was by no means the only one. Celluloid horn was also made, requiring, like ivory, special processing to produce horn-like striations and marbling. Tortoiseshell was one of celluloid's most important imitations, in part because it could be done so well and also because natural tortoiseshell was rapidly becoming quite scarce. Numerous methods were used to make imitation pearl and mother-of-pearl from celluloid, as well as coral and amber imitations, the latter being widely used for tobacco-pipe stems despite the flammability of the material. Marble and onyx could also be mimicked by celluloid, as could almost any other decorative mineral. The production of beautiful effects with celluloid, almost all of them imitative, was the subject of innumerable experiments and patents in the last quarter of the nineteenth century. The result of these efforts, however, was often less to prepare celluloid for the direct substitution of the imitated material in some particular use than it was to give the plastic a familiar and readily acceptable appearance regardless of the object that was actually made from it. Hence celluloid visiting cards looking like ivory or tortoiseshell were not intended so much to suggest real ivory or shell card — such things did not exist — but rather celluloid cards in an indisputably familiar form.

To make celluloid a success, the Hyatts had not only to determine what products were saleable and what forms were acceptable, they also had to establish the basic price identification of their product. This was something that Alexander Parkes and his partners had been unable to do. Even without complete knowledge of how important camphor was in making a good pyroxyline plastic, Parkes was able to produce a useful material if care and skill went into its manufacture. But there was early pressure to produce parkesine at 'a shilling a pound', and hence lower quality materials and less care went into the product, resulting in a largely worthless substance. The makers of celluloid, despite their superior technical knowledge, could have made the same mistake if they had insisted on trying to make a very cheap material. This was not their object, however, as most clearly shown by the effort to market celluloid in competition with rubber for dental plates. The difficulty that the material experienced in this market showed the Hyatts and their partners that celluloid would generally have to be price-competitive. Therefore, the material was established as an intermediate substance—never as cheap as rubber and a few other materials, but definitely cheaper than the natural materials that it imitated so well. Here too celluloid set a precedent for the modern plastics — rarely the cheapest alternative (pound for pound), but cheap enough to provide an economical substitute for many traditional substances.

The success of the Celluloid Manufacturing Company after it set up its plant in Newark was far from immediate. Indeed, in its first years the company sustained substantial losses, totaling more than $115,000 between 1873 and 1876. This was due not only to initial difficulty in selling their product, but also to a sizeable fire in 1875 and to the normally high capitalization costs that would be expected in the start-up of any sophisticated chemical industry. After this difficult beginning, however, the company became an eminently profitable operation. The marketing strategy pursued through the licensed companies combined with substantial economies in manufacturing to put the company on a firm footing. Between 1876 and 1879 the company was able to halve the average cost of making a pound of celluloid, from $2.06 to $1.03. The average price paid by consumers fell relatively little during this same period, from $1.939 to $1.366 per pound. The simultaneous tripling of volume in this period (to 212,921 pounds produced in 1879) gave the

Newark firm a healthy financial position.[25] Indeed, for fourteen years after 1876, until the company was reorganized in 1890, the annual dividends paid to stockholders averaged 34%.[26] Clearly, by the end of its first decade in business, the Celluloid Manufacturing Company had succeeded in putting itself on a firm financial footing and had given its product a solid technical and economic foundation.[27]

The successful introduction of celluloid was not an easy accomplishment. The problems of the makers of parkesine illustrated the necessity for the solution of both technical and commercial problems before a product like celluloid could be viable. The difficulties of the Hyatts with billiard balls and dental plates seem to have taught them the importance of recognizing both the technological and the economic limiations of their invention. To make celluloid into a success it was necessary not only to discover the chemical and mechanical means to make a stable and workable pyroxyline plastic, but also to determine the useful properties of such a material. Celluloid did not have the density of ivory nor the durability of hard rubber, but it did have a versatility of colouring and ease of working that was matchless. Celluloid could never be as cheap as rubber or horn, but it could be made much more attractive than them. It was in terms of traditional materials and traditional products that celluloid had to be defined, and the success of the Hyatts was in finding an economically viable definition.

Notes

1. The only attempt at an extended history of early plastics is Morris Kaufman's *The First Century of Plastics; Celluloid and its sequel* (London: The Plastics Institute, 1963). Kaufman's work, however, was directed towards a broad audience as a part of the Plastics Institute's celebration of plastic's centennial, and it lacks the appurtenances of a scholarly work.

2. Edward C. Worden, *Nitrocellulose Industry*, 2 Vols. (New York: D. Van Nostrand, 1911), 2:815.

3. *Ibid.*, 2:828; Robert Taft, *Photography and the American Scene* (New York: Macmillan, 1938; reprint ed., New York: Dover Publications, 1964), pp. 118–19.

4. 'Death of Mr. Alexander Parkes', *Engineering* 50 (25 July 1890): 111; Kaufman, *The First Century of Plastics*, p.17.

5. London International Exhibition of 1862, *The Illustrated Catalogue of the Industrial Department — British Division* (London: for Her Majesty's Commissioners, 1862[?]), p. 103.

6. Kaufman, *The First Century of Plastics*, p. 25.

7. Alexander Parkes, 'On the Properties of Parkesine and its Application to the Arts and Manufactures', *Jour. Soc. of Arts* 14 (22 December 1865): 82.

8. 'The Parkesine Company, Ltd.', exhibit in Spill v. Celluloid Manufacturing Company, Circuit Court, Southern District of New York, 1880. In 1875 Daniel Spill brought suit against the Celluloid Manufacturing Company of Newark, N.J. for infringement of certain U.S. patents which he had taken out on parkesine-like preparations. The suit remained in litigation for fourteen years, ending in dismissal of Spill's appeal by the U.S. Supreme Court in 1890 due to Daniel Spill's death three years earlier. While Spill did have some claim to having made technical improvements in Parkes' process, most evidence suggests he no more recognized the importance of camphor in producing a successful plastic from nitrocellulose (see below) than had Parkes. A fundamental part of the defence of the

Celluloid Manufacturing Company was Parkes' anticipation of all that was useful in Spill's patents, and Alexander Parkes himself testified as a witness for the defense. The records of the case are an invaluable source for studying the early history of celluloid. Most of the records are to be found in a printed transcript prepared for Spill's appeal to the U.S. Supreme Court after his case was finally rejected by the Circuit Court. This transcript can be found filed under case number 12305 in the 'U.S. Supreme Court Appellate Case Files', in Record Group 267, National Archives, Washington, D.C. Depositions and exhibits taken in London for the trial can be found in Record Group 21, 'Old Equity Case Files, 1846–1877' (Docket number 7-336), in the Washington National Records Center, Suitland, Maryland.

9. Alexander Parkes, Spill v. Celluloid Manufacturing Company.

10. Alexander Parkes, *Brief Account of the Invention and Manufacture of Parkesine* (Birmingham, Corns & Bartleet, 1867), p.3.

11. Descriptions of celluloid manufacture may be found in Worden, *Nitrocellulose Industry*, 2:582–614; The Celluloid Company, *Celluloid. How Celluloid is Made* (Newark, N.J. [?]: The Celluloid Company, n.d.); and René Dhommée, 'Fabrication du Celluloïd', *La Revue Technique* 26 (10 May 1905): 372-3.

12. John W. Hyatt, 'Address of Acceptance' (of the Perkin Medal), *Journal of Industrial and Engineering Chemistry* 6 (February 1914): 158.

13. *Ibid.*, pp. 158–9.

14. James H. Prothero, *Prosthetic Dentistry* (Chicago: Medico-Dental Publ. Co., 1916), p.554.

15. *The Dental Cosmos* 13 (1871): unnumbered (advertising) page.

16. A nineteenth-century rubber plate *sans* teeth from the Smithsonian Institution collections was weighed to determine this.

17. *The Dental Cosmos* 13 (1871): unnumbered (advertising) page.

18. Jean Delorme and Pierre Laroux, *Les Conquêtes des Matières Plastiques en Mèdecine et en Chirurgie* (Casablanca: Les Editions Amphora, 1950), p.88.

19. Prothero, *Prosthetic Dentistry*, p.439.

20. Carl Marx, ('The Rise of the First Great Plastic Industry', *Plastics* 4 (December 1928): 670–1.

21. 'Master's Report', in Spill v. Celluloid Manufacturing Company, pp. 998–1006 (see note 8). The first decision handed down in Spill's suit was against the Celluloid Manufacturing Company. Before the company's appeal was accepted, the Court appointed a Master to investigate the profitability of the Newark company in order to determine damages. This 'Master's Report' is the only surviving record of the Celluloid Manufacturing Company's first years in business.

22. William F. Ford, *The Industrial Interests of Newark, N.J.* (New York: Van Arsdale and Co., 1874), pp.21–3.

23. Great Britain, Home Department, *Report of the Departmental Committee on the Use of Celluloid in the Manufacture of Toys, Fancy Goods, etc.* (London: HMSO, 1938), pp.3–4.

24. Celluloid Manufacturing Company, 'Licenses, Contracts and Patents 1872 – 1878, 2 vols., unpublished ledger held by Celanese Plastics Company, Summit, N.J.

25. 'Master's Report', in Spill v. Celluloid Manufacturing Company, pp.1062–3.

26. Marx, 'The Rise of the First Great Plastic Industry', pp.671 & 684.

27. The United States *Census of Manufactures* for 1880 included the new category, 'Celluloid and Celluloid Goods'. Under this heading it reported five establishments in the state of New Jersey and one in the state of New York, producing products valued at $1,261.540 for the census year.

Iron Arched Bridge Designs in Pre-Revolutionary France

J.G. JAMES

Whenever Frenchmen relinquish their fiddling and dancing and cultivate the art of iron-making &c, England will tremble.

William Wilkinson, c.1777

1. Introduction

In 1769 an ornamental bridge, undoubtedly a light wrought-iron structure, was built in the grounds of Kirklees Hall near Huddersfield by a Leeds smith named Maurice Tobin. Its construction, although reported in the local newspaper, appears to have aroused little interest.[1] Ten years later the massive cast-iron arch erected at Coalbrookdale after several years of planning and debate achieved instant international fame.

In Britain the 'Iron Bridge' *par excellence* had no immediate successors and a decade elapsed before the next serious attempt. In France on the other hand much discussion of iron bridges occurred in the years between the passage of the Coalbrookdale Act in 1776 and the outbreak of the Revolution in 1789, and several proposals for such bridges were put forward. Since these French plans were certainly known in England too, a consideration of them should necessarily precede any study of British iron bridge designs from 1790 and later.

At first sight it may seem surprising that France, with its world-renowned Corps des Ponts et Chaussées, should in this new technique of bridge-building have been eclipsed by Britain; but two relevant factors may be noted. Firstly, the French system was a bureaucratic one. The very existence of an official corps of bridge-builders was an impediment to innovation in design. Any new design was invariably sent before a committee for evaluation, and there the senior officers who were wedded to the traditions of masonry and timber, would insist on the risks and difficulties inherent in novelty. As one promoter of an iron bridge, Léonard Racle, wrote in exasperation: how on earth did the first *masonry* arch ever get built? Secondly, the French iron industry had progressed less than that of Britain during the eighteenth century. Formerly French iron-working skill had been famous — and many English technical terms related to iron had been borrowed from French. For example, cast iron had first been used for large water-pipes at Versailles in 1672 (Belidor's date); then, owing to a reaction against the incautious use of such pipes inadequately buried in city streets where their brittleness caused fractures, cast-iron production had been generally neglected. Réaumur's toughening process,[2] intended to make cast iron malleable, was like Prince Rupert's earlier project in England only of very limited use. Thus by the 1770s the use of so unreliable a material as cast iron for engineering purposes was unacceptable to the engineers of the Ponts et Chaussées. Interest in British iron-industry innovations brought about the introduction of new smelting and founding techniques at Le Creusot in the next decade, but too late for much to be accomplished before the Revolution.

Because of the French mistrust of cast iron the designs for most of the bridges discussed in this paper employed wrought iron, a material in whose use the French excelled. Wrought-iron reinforcements had long been employed in masonry, and light beams of the same material were used for flooring from 1782 onwards. Nevertheless, there was a formidable problem to be overcome in building up the necessarily large ribs of bridges of long span, intended to carry vehicles, from the available small bars and billets rarely weighing more than a hundred pounds. At the end of the eighteenth century the largest forgings were warship anchors of about five tons, each of which took months to make and was valued at £500. Contrast that expense with the half-rib of the Coalbrookdale arch (about $70' \times 9'' \times 6''$) of about the same weight but costing (perhaps) only about £50–60. Would-be iron bridge builders in France clearly experienced different constraints from those prevailing in Britain, and their primary problem was to build a strong frame from relatively small forgings.

In timber construction well-known precedents existed. In his *Nouvelles inventions* (1561) Philbert de l'Orme (1518–77) had described a vertically-laminated arch rib made from short, curved planks having staggered radial joints. Alternatively, an arched rib could be constructed from voussoirs of framed timber, assembled like the stone voussoirs of a masonry arch. The origin of this idea may be found in one of Palladio's designs, widely used for eighteenth century ornamental garden bridges, but a true timber-voussoir design was published in 1700, among a collection of inventions attributed to Claude Perrault (1613–88).[3] Perrault, a scholar as well as a successful architect in practice (who used reinforcing iron in the Louvre), proposed for a bridge that was to cross the Seine at Sèvres two $180'$ arches, each of 5 ribs; each rib was to be formed of 17 voussoirs framed from $12''$ timber, graduated in depth from abutment to crown. In a popular book J. E. Montucla (1725–99), the historian of mathematics, showed an alternative design for a $100'$ span made up of $5'$ voussoirs in 1778, so the idea was not forgotten.[4]

Perrault's design included a further influential feature in that the ribs were splayed in plan towards the abutments, in order to increase the lateral stability of the bridge and reduce any tendency to warp; the arches were to be $36'$ wide at the ends but only $18'$ wide at the centre.

In July 1771 another form of timber-arch bridge design was also studied by members of the Paris Académie Royale d'Architecture, for whom copies of the designs were specially prepared: this was employed in the Grubenmanns' bridge (some 200′ in span) over the Limmat at Wettingen, Switzerland, built in 1764–6. Whereas in earlier Grubenmann bridges straining-beams were used in a cumbersome (though effective) way, in the Limmat bridge they were replaced by an arch of curved timbers intended to act, with the deck suspended below it, like a bow and string.[5]

2.   French Claims to Very Early Iron Bridge Designs

French writers have recorded four plans for iron bridges before 1760, but no contemporary evidence for them has yet been produced. The earliest reference traced by me dates back to 1779, the year of the Coalbrookdale bridge. In later years, amid Napoleonic Anglophobia, exaggerated claims were circulated.[6]

As enumerated concisely by Vincent de Montpetit (whose career will be considered in detail later) in his *Prospectus* of 1783 the four early proposals were as follows:

> In this century Dr. Desaguliers conceived it [the use of iron] for the Thames; Mr. Garrin was on the point of executing one at Lyon in 1719; later another was proposed for the St. Vincent Bridge over the Saône; and in 1755 Messrs. Goiffon and Montpetit proposed it [iron] for crossing the Rhône.

Clearly all but the first of these proposals relate to the city of Lyon. Considering them in turn:

a) *John Theophilus Desaguliers* (1683–1744) was only French by birth. Brought as an infant to England by Huguenot parents, he became a lecturer in experimental philosophy at Oxford and London. His *Course of experimental philosophy* (1734, 1744) was widely read, and contains much on technology. He was consulted during the protracted discussions before the building of Westminster Bridge, for which the first Act of Parliament was passed in 1736, but nothing connecting Desaguliers with any proposal for an iron bridge has been discovered.

b) *Garrin (or Guérin)* is a wholly unknown person. His bridge is first mentioned in a report by Jean-Rodolphe Perronet (1708–94) on an iron-bridge proposal put forward in 1779. After agreeing that iron might be of service for bridges in districts where stone for masonry was not available, Perronet wrote that iron need not be employed for 'arches of medium span such as the three which it was proposed to construct of iron over the Saône at Lyon about 60 years ago [1719], each of 76′ span; where, however, economic reasons caused a wooden bridge to be preferred even though one of the arches had been prepared already and assembled in the works or on adjacent ground'. Perronet did not name the designer.

At this point one cannot omit the almost wholly erroneous paragraph penned by Émiland Marie Gauthey (1732–1806) whose papers on bridges were published posthumously by his nephew C.L.M.H. Navier (1785–1836). In the section on iron bridges (1813) he wrote:

> 'The idea of using iron in bridge construction is very old and information concerning it can be found in 17th century Italian works. Desaguliers revived it in 1719 [!] and about 1755 an iron bridge of three arches of 25m span was undertaken at Lyon; one of these spans was erected in the works but the structure was not completed for economic reasons and a wooden bridge replaced it.[7]

Obviously Gauthey miscopies from Perronet and Montpetit. Gauthey also gave the false date of 1782 (for 1779) for Montpetit's own iron-bridge proposal on which Perronet had commented, introducing yet another error into the *Encyclopédie méthodique* (1825),[8] where the Saône Bridge at Lyon that was abandoned is dated 1722. A third piece of misreporting occurs in the classic work on masonry bridges by Dartein;[9] quoting the Perronet report of 1779 with its allusion to a Lyon proposal of 60 years before, Dartein interpolated the words 'proposed by M. Genin' presumably taken from Montpetit, with Garrin corrupted into Genin. Racle gave the name as Guérin.

c) *The anonymous designer* of the Pont St Vincent at Lyon may have been Goiffon, a shadowy figure who died before iron bridges were of real interest in France. He was an early collaborator of Vincent de Montpetit.

At Lyon the river Saône is generally 120–180 m wide,[10] though at one point on a bend it narrows to about 90 m. Here is the Pont St Vincent. According to the modern historian Guillemain[11] there was an ancient wooden bridge on this site, restored many times between 1632 and 1711, then totally rebuilt in wood, the designer being named Aubert. It is generally held that this new bridge is the one depicted in Plate 17 of Gautier's treatise of 1716 as a three-span bridge with trusses of 12, 15 and 12 toises.[12]

The wooden bridge of 1711 might be expected to last without trouble for 20 years of more; therefore any plan for an iron bridge would be later, probably, than 1730. If there was one, nothing came of it. Some form of wooden bridge continued on this site well into the nineteenth century.

In a Lyon newspaper of 1807 a writer signing himself 'N' made further allusions to this plan, in a thoroughly chauvinistic spirit.[13] He wished to restore to France the honour of inventing iron bridges which the enemy (England) had usurped. Accordingly, he claimed that the plan for the first such bridge in Europe, 'perhaps in the world', was conceived by an artist 'a painter from Lyon . . . in the middle of the last century' for the Pont St Vincent. In the library of an architect named Basfer he had seen the design, a single-span arch of 254′ × 18′. Later, however, he speaks of '*two* artists of this city' and since both Goiffon and Vincent de Montpetit were artists it seems clear that 'N' was referring to them, but possibly confusing the present proposal with the next, d).

d) *Goiffon and Vincent de Montpetit* can have put forward no more than a tentative idea for an iron bridge to cross the Rhône at Lyon in 1755. The only references to it are that of Vincent de Montpetit already quoted and similar later comments by him to the effect that this 1755 plan foreshadowed his designs of the 1770s and 1780s.

At Lyon the Rhône is from 205 to 280 m wide; it was then crossed only by the multi-arch masonry bridge known as the Pont de la Guillotière, built originally in the thirteenth century.

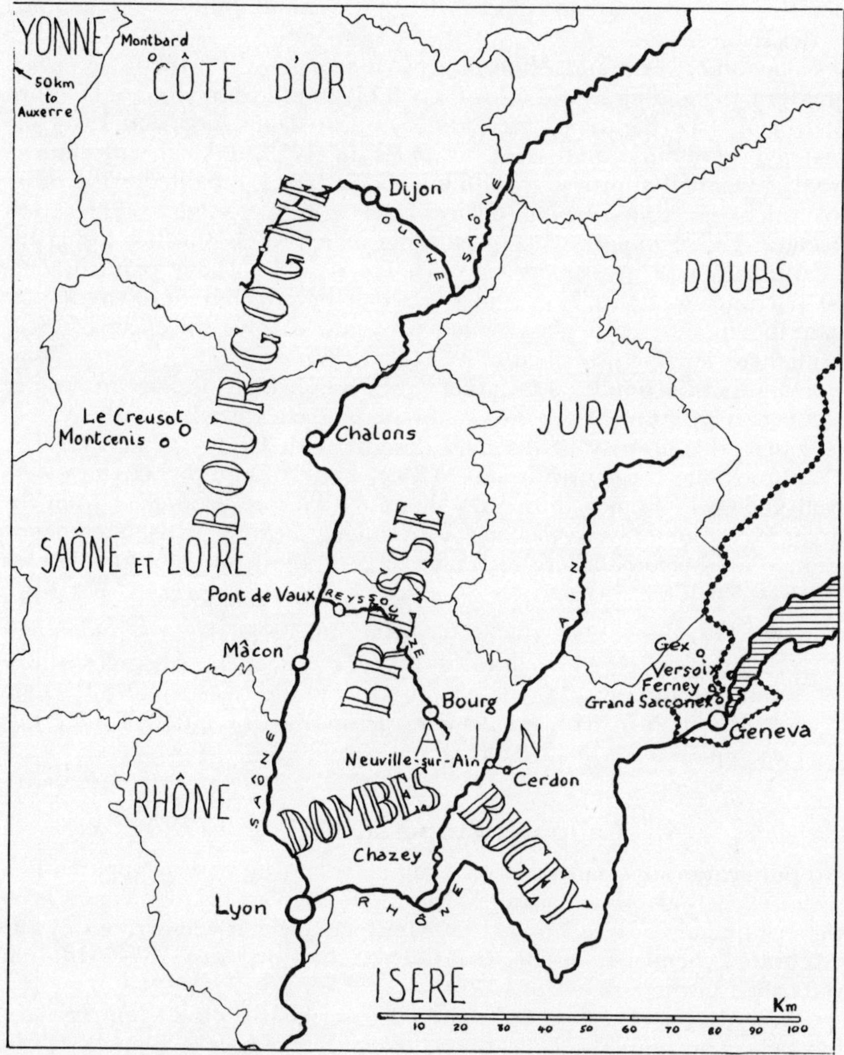

Figure 1. Map of south-eastern France marking the places named in the text (based on V. A. Malte-Brun, *Atlas de la France Illustrée*).

3.   Notes On Goiffon and Vincent De Montpetit

Several authors named Goiffon wrote in the eighteenth century, and the 'bridge-designer' was probably Georges Claude Goiffon (1712–76), a native of Lyon, described as painter, architect and teacher. He died either at Cesson,[14] or Alfort,[15] near Paris, having published (it is supposed) in 1768 on the *Hippomètre* (a device for measuring horses for artistic purposes) and in 1772 on *L'art du maçon-piseur*. More definitely, Goiffon and Vincent published a work on the representation of animals in painting and sculpture at Alfort in 1779; presumably these were the men of the Lyon bridge.

The other, Armand Vincent de Montpetit (1713–1800), achieved greater fame and figures in most French biographical dictionaries.[16] Born at Mâcon (see Figure 1), well-to-do, educated at Dijon and Lyon, he married at Bourg and went to Paris in 1753, taking with him a newly-invented improved pendulum clock. He is credited with other inventions in time-keeping devices and a new plough, besides iron bridges. He seemingly took to painting as a profession after losing his wealth in a bad investment in 1763. He is said to have painted some 40 portraits of Louis XV, and devised a new technique of miniature painting in oils upon glass which was said to give an effect of great brilliance. This he described in print in 1775.

Montpetit's public association with iron-bridge designs in France seems to have sprung from the Coalbrookdale scheme, which induced him to assert the priority of his own design (with Goiffon) of 1755. The Coalbrookdale committee began to meet in 1775, and its activities were well known in France through William Wilkinson, brother of John the great iron-founder, who was busy introducing the new British methods to France at this time. Here again we may quote from Montpetit's own *Prospectus* of 1783:

> In 1777 and 1778 two projects of this type [that is, for iron-bridges] appeared, each on a different system, by Messrs. Calippe and Montpetit. M. de Morveau of the Académie of Dijon reviewed them with judicious criticism and the latter of these two authors replied in 1779.

4.   Calippe's Proposal, 1777 or 1778

No publication by Jean-François Calippe is known; he was apparently a *serrurier*, or ironwork fabricator.[17] Information about his plan comes from the critique already mentioned by Montpetit, which was written by the celebrated chemist Louis Bernard Guyton de Morveau (1737–1816) and published in the autumn of 1779.[18]

According to Guyton, Calippe had described it 'on 20 January last' [therefore, presumably, 1779] to the assembly at M. de la Blanchérie's in Paris; which assertion does not prove Montpetit's date incorrect. Calippe's bridge was to be of 600′ span, resting on masonary abutments without lateral thrust. There is no drawing, but Guyton's verbal description indicates a bow-string arched girder, the main members being composed of wrought-iron plates on edge with the upper and lower chords connected by vertical rods and diagonal braces. Guyton adds that cannon-balls were in some way to be hung along the bridge, in order to damp down live-load oscillations which might be created in so light a structure; he was doubtful of the value of this feature. We may conjecture that Guyton quite misunderstood Calippe's drawing (now lost): that the cannon-balls were drawn (or shown on a model) merely to demonstrate the use of the diagonal bracing in spreading point-loads applied to the deck (compare the drawings of the braced bow-string girder in the book published by Robert Fulton some years later).[19] Guyton also gave his opinion that a 600′ span was excessive: it would be cheaper and better to provide a central pier and halve the length.

Although Calippe's plan could not have been practised in the 1770s because of the difficulty of making and erecting such a bridge, he should perhaps be recognized as the originator of iron bow-string designs. It is possible that the relatively small wrought-iron bow string girders proposed by Ango, another serrurier, in 1782[20] and widely used in buildings thereafter may have derived from Calippe's original idea. Calippe presented his design to the Académie Royale d'Architecture in 1784 but then he disappears.

5.   The Proposals Of Guyton de Morveau, 1779

Before considering the more extensive work of Vincent de Montpetit, it is convenient to mention the chronologically prior proposals of Guyton de Morveau.

Guyton's own ideas on iron-bridge construction emerge in his 1779 paper[21] by way of comments on two designs which he had come across in Paris during the previous winter (1778/9); he asserted that he had earlier thought of using iron for bridges but chose this method of disseminating his ideas in order to avoid any possible charge of plagiarism. The first of these designs was Calippe's, just noted; the second was by Montpetit for an arch of 200–250′ span (see section 6).

After describing these plans, Guyton describes his own for a bridge of 200′ span, of wrought iron, with a rise of 40′. It was to have 26 arched ribs, each 220′ long, 9″ deep by 2″ thick; the ribs spaced about 1′ apart made a bridge 30′ wide at the centre and 36′ wide at the abutments. Like Perrault Guyton preferred to splay the haunch to obtain a buttressing effect, rather than keep the ribs equidistant. He proposed to laminate them from three sets of plates each $\frac{3}{4}$″ thick, using manageable lengths with staggered joints in the manner of de l'Orme. The two middle ribs were to be of double depth and the two outer ones of treble. To avoid weakening the plates with bolt holes he proposed to clamp them together. Laterally the ribs

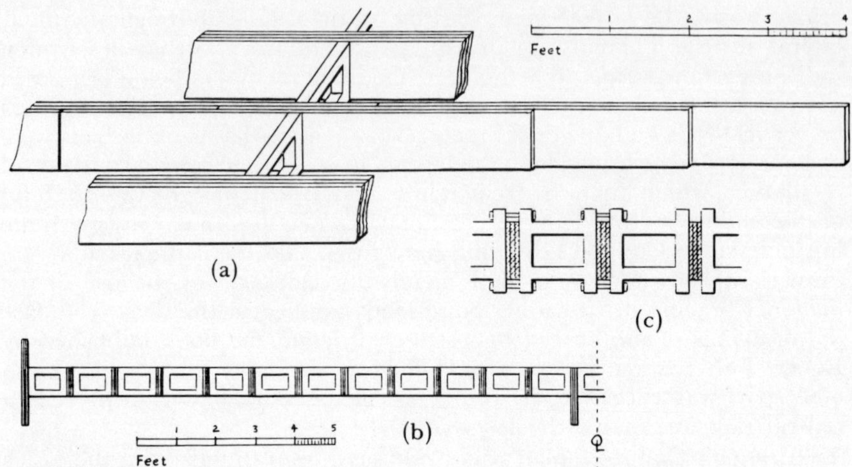

Figure 2. Reconstruction of Guyton de Morveau's rib-system 1779: (a) perspective view, (b) half section, (c) possible method of clamping the laminae.

were to be connected by frames made of 2″ square bar, set at 6′ intervals (Figure 2). The weight of this whole structure he calculated at under 600,000 pounds — 506,780 for the ribs and 67,968 for the frames. To make the iron 'absolutely unrustable' it was to be quenched in oil 'as demonstrated by M. de la Folie'.

6.   Vincent de Montpetit's Proposals, 1778–88

In 1778/9 Montpetit made a model bridge and wrote three papers which were not printed and do not survive. The evidence for his work consists of the comments of Guyton de Morveau[18] (who had met Montpetit), Montpetit's published reply to Guyton,[22] a report on his plans by Perronet,[23] and papers published by Montpetit in 1783[24] and 1788[25], which refers to the earlier work. From these sources a clear picture of the progress of his ideas emerges.

(a)   FIRST 'DESIGN PROSPECTUS', 1778

In his 1783 *Prospectus* Montpetit recorded that his first design — after the collaboration with Goiffon, that is — had been made in 1778, following Calippe. Further, in replying to Guyton in 1779 Montpetit declared that when he and Guyton had met (in the winter of 1778/9, wrote Guyton) in the presence of the Prince de Condé, the Prince 'had asked for a further look at the Design-prospectus of my plan . . .'. From this it would appear that Montpetit had something on paper at that time.

According to Guyton de Morveau (who had, however, seen neither a detailed paper by Montpetit nor a model bridge) the design was for 'a very shallow arch' composed of 'several iron plates on edge'. He understood that for economy the plates were to be made of cast iron, an expedient of which he disapproved.

(b) MODEL AND MEMOIR: JULY 1779

Montpetit stated in his 1783 *Prospectus* that he had presented a model and a memoir to the Académie des Sciences in 1779, which model was:

> exhibited for four months in the Assembly Room in order to obtain comments on methods of improving it from the Academicians. It was then placed during the following winter (1779/80) in M. de la Blancherie's Concourse of Arts and Sciences for criticism, not only by national scientific and technical experts but also by foreign savants attending these assemblies.

Presumably the voussoirs in this 1779 model were made of wrought iron; they were strapped together at top and bottom by wrought-iron bands. So, at any rate, Montpetit stated in his 1788 paper. There Montpetit also affirmed that the voussoirs in the full-size bridge were to be made of wrought iron, but Perronet confirms Guyton's statement that in 1779 Montpetit intended to make them of cast iron.

For in his report Perronet wrote:

> According to the first designs submitted by M. de Maupetit [*sic*] with his memoir of July 1779 the arch is to be made with cast-iron frames in the shape of voussoirs each 8 ft long and 6 ft high, the lower members of which are to be 10 inches deep, the upper ones 12 inches, and the vertical bars 6 inches, all 1½ inches

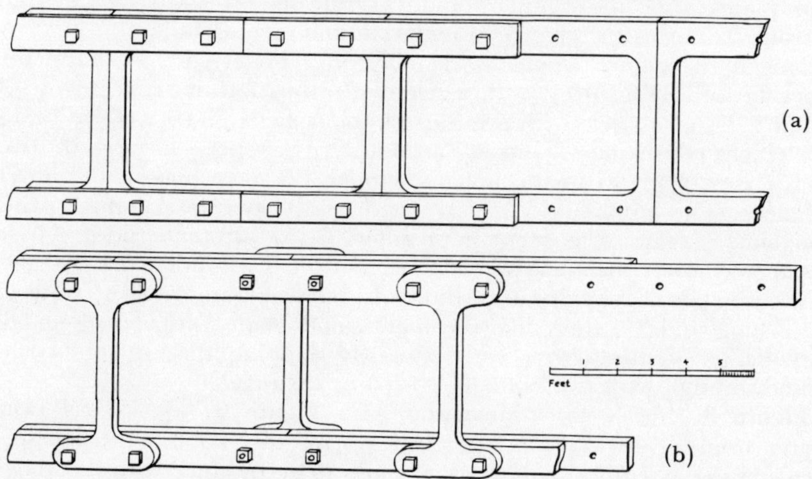

Figure 3. Reconstruction of Montpetit's rib-systems: (a) 1778 design, with cast-iron voussoirs and wrought-iron straps (from Perronet's description), (b) 1779 design, with wrought-iron laminated ribs (from Montpetit's drawing).

thick: these frames are to carry alternately on their abutting sides mortices and tenons to prevent them from sliding. The top and bottom members are to be overlapped on each side by bands of wrought iron of similar dimensions, arranged with lapped joints and each fastened to the iron frames by four bolts. 25 such frames or voussoirs are to be used for each of the 13 curved ribs of which the arch is to be composed, connected laterally by horizontal and diagonal iron bars, and covered over with a deck, footpaths and railings, all in iron, to carry the roadway over the arch.

Figure 3(a) has been drawn to illustrate this system.

(c)   NEW DESIGN AND SUPPLEMENTARY MEMOIR: AUTUMN 1779

In replying to Guyton de Morveau's comments Montpetit acknowledged that he had planned arch ribs composed of 'hollow' cast-iron voussoirs, to which wrought-iron plates were to be bolted, and in justification of such a use of cast iron instanced the successful Coalbrookdale bridge, besides making reference to Réaumur's toughening process.[26] However, realizing by now the strength of the official distrust of cast iron, he strove in this reply to emphasize that he had always contemplated making voussoirs of wrought iron also, and he now concentrated on this form of iron in his new memoir addressed to the Academy. This 'supplement addressed to the Commission nominated to report on the [original] plan' appears to date from the autumn of 1779. The new design (which, he claimed in 1783, was to be 'simpler, lighter and cheaper yet just as strong') was also for a 200′ arch, of similar outline, the ribs being made up wholly from wrought-iron plates after the manner of de l'Orme as proposed by Guyton: Perronet confirms the change of material, and gives details.

This arch was to be 40′ wide (Perronet preferred 43′ abutments tapering to a 37′ centre), with a rise of 12′ thus forming a segment (of about 27½°) of a circle with a radius of about 422′. There were to be 13 ribs each consisting of upper and lower members formed of two thicknesses of plates in the upper member 12″ deep and in the lower 10″ deep. The individual plates were to be 10′ long by 1½″ thick, with the joints staggered to occur at 5′ intervals on alternate sides. These upper and lower members were to be connected by radial tie-bars also located on alternate sides at 5′ intervals. The tie-bars (moises) were to be 6′ long and 1½″ thick like the plates, and made with a long shank 6″ wide terminating in 3′ wide *pattes* overlapping the rib joints and bolted through with two bolts (8″ × 2⅔″) at each end.

Figure 3 (b) — for comparison with Figure 3 (a) — has been made from a drawing in the Archives of the Ponts et Chaussées showing the assembly of the various parts in accordance with the above description. Although this drawing has a vague title and lacks date and signature, it clearly belongs with Montpetit's papers of 1779.[27]

As in his earlier design, the ribs were to be interconnected laterally by means of horizontal and diagonal bars. The deck was to be of iron plates 1″ thick; the footpaths and railings also of iron are not described in detail.

Perronet calculated the weight and cost of such a bridge at 1,882,972 (French) pounds and 657,936 *livres* (the latter figure is incorrect and should read 757,936):

	(French) pounds	Livres
Plate iron for ribs	1,186,640	
Bar iron for lateral frames	186,300	
Deck plates	356,666	
Footpaths, railings, etc.	153,366	
Total weight of structural iron thus	1,882,972	
Cost at 5 *sols*/pound (including erection)		470,743
2,080 bolts (8″ × 2⅔″, each with one nut) weighing 18 pounds each		
Total weight of nuts and bolts	37,440	
Cost at 8 *sols*/pound		14,976
Fitting bolts, reaming holes, etc., at an estimated cost of 1 livre per bolt		2,080
Cost of scaffolding, wooden centering, machinery and erection		60,000
Abutments, wingwalls and cofferdams		120,000
Paving of bridge and approaches (about 355 square *toises* at 15 livres)		5,325
Terraces at each side of the bridge		5,000
Painting, rust-proofing and contingencies		20,000
10% for direction and profit		59,812
Total cost		657,936

Perronet concluded that such a bridge was ingenious, simple and feasible but several points troubled him, notably the efficiency of the rust-proofing and the excessive stress-concentration at the bolts when the bridge expanded and contracted. Chiefly, however, he disapproved of the design on economic grounds, devoting most of his report to rather biassed computations showing that an equivalent masonry bridge would be as costly but stronger, while a wooden one would be a great deal cheaper. For these comparisons, Perronet first considered gritstone masonry (from a quarry near Fontainebleau) with a crushing strength of 6,245 p.s.i;[28] the 600′ masonry bridge, 45′ wide, at Neuilly-sur-Seine had cost 2,394,900 *livres* and therefore an equally long bridge 40′ wide would (he reckoned) have cost 2,128,800 *livres*, about the same as three of Montpetit's spans. Secondly, for a timber bridge using oak (7,200 p.s.i. in compression) he calculated the cost at a third only of this sum. Admittedly, timber was less durable but (Perronet argued) if the main members were sheathed with lead, as he had treated them at St Cloud, they might well last for a century. In fact, the St Cloud bridge — 22 years old — was soon to require costly maintenance and barely lasted 50 years.

Perronet had still a further objection: Montpetit's iron was too weak. Since the strength of wrought iron in compression was too great for measurement, he took its tensile strength (estimated at the equivalent of 57,600 p.s.i.) as a guide. Arguing that the lower member in each rib might be in tension at the crown, he merely considered the upper member and took the rib cross-section at 36 square inches. He then concluded that, taking a safety-factor of four, the rib was too weak by more than one-half.

Montpetit understood from Perronet's report that the Commission would be unable to reach a decision about his proposed iron bridge without undertaking tests. He therefore withdrew his memoir, in order to submit in its place more formal revised proposals embracing a programme of tests. This document was to be his *Prospectus* of 1783.

(d)   MONTPETIT'S *PROSPECTUS*: 1783

From its title the *Prospectus* is concerned with a bridge of a span varying between 120 and 600 feet.[24] It has a large folding plate, often reproduced, showing an 'Iron bridge with a single arch of 400 ft span, fitted with a pyrometer showing the extent of expansion and contraction of the iron with air temperature' (Figure 4). This bridge bears little resemblance to that described in the text: it was to consist of twin-membered ribs, about 7′ deep at the crown and about 15′ at the sides, with the pyrometer shown prominently at the crown.[29] The rise is about 35′, with the main members (upper 15″ and lower 18″ deep) connected by oddly shaped radial struts. The roadway resting on the upper members is objectionably steep. A second deck for pedestrians rests on the lower members. The weight of iron was put at 1,800,000 pounds.

After reviewing the history of his bridges, Montpetit put forward the advantages of using single large spans for crossing busy rivers, and

Figure 4. 400-foot bridge drawing from Montpetit's *Prospectus*, 1783: left-hand half of the view in elevation.

then calculated the weight and cost of a 300′ span iron arch similar to the 200′ bridge already examined by Perronet, with a rise therefore of 18′: to cheapen it Montpetit proposed to omit the footpaths at the side and reduce the width from 40′ to 28′. He also omitted deck and railings from his calculations, suggesting that these could be wooden. The 9 ribs were now uniformly 6′ deep, made of two members each formed of a pair of overlapping $12'' \times \frac{2}{3}''$ laminae. These were connected by uprights with shanks $48'' \times 5'' \times 1''$ and feet $12'' \times 6\frac{2}{3}''$. The total weight of iron he puts at less than 600,000 pounds, and the cost at less than 300,000 *livres*, including an allowance for rust-proofing. Should it be desired to widen the bridge extra galleries could be added for about 30,000 *livres* each.

Montpetit also proposed that full-scale rib components (each $84'' \times 8'' \times \frac{2}{3}''$) should be tested in a special machine under compression, impact, and vibration, and other parts similarly tested under tension. The effect of thermal expansion on the abutment thrust was also to be explored. Should the bridge never be built the testing-machines and the results recorded would nevertheless remain of great value.

(e) PAPER IN THE *JOURNAL DE PHYSIQUE:* 1788

Montpetit now vanishes from the scene for five years: evidently no money for tests or trials was forthcoming. Then in 1788 he again published on bridge-design, this time surely stimulated by the fresh interest in iron that resulted from Thomas Paine's visit to Paris with new proposals in 1787.[30]

Montpetit's new paper traverses much familiar ground.[25] Because iron was both strong and expensive it was natural to use as little of it as possible in a structure, with the consequence that vibration and wear at the joints might prove excessive. Thus the structural members must be so disposed that the maximum inertia should be built into the system. Thermal expansion and contraction also had to be understood and provided for. Recalling his own experience of 'more than 30 years' Montpetit claimed that since 1779 his experiments with special machines had yielded a quantity of papers, plans and models of great value for the design of long-span iron bridges 'whether in compression, extension, suspension or all three together'.

He then turned to yet another account of his wrought-iron voussoir principle. He explained how the outline of the voussoirs could easily be obtained by imagining a 'platte-band' consisting of an 'assembly of parallelograms' stretched across the river, with two arcs inscribed on it to represent the intrados and extrados of the proposed arch. The longitudinal members in each voussoir-frame would be made stronger than the radial members, and· (to increase stability) the frames could be coupled by additional bands of wrought iron, with joints staggered with respect to those in the voussoirs. These principles, he claimed, had been exhibited in a model at the Académie des Sciences 'for the last ten years'.

A similar arch could be built up from wrought-iron plate, rather than forged voussoirs, as he had described in his *Prospectus*.[24]

By now Montpetit was 75 years old, and this was his parting shot. He must have found some champions since, after the Revolution, in 1793, he was awarded a gratuity of 8,000 francs in respect of his various inventions. He died in 1800, before France had constructed a major iron bridge. The fate of his papers and models is unknown.

7.   Léonard Racle's Proposals: 1782–90

Léonard Racle (1736–91), born at Dijon, formed his first plan for an iron bridge in 1782. He is said to have been the pupil of a little-known Burgundian architect-engineer named Montin (or Moutin) de Saint-André, who unsuccessfully offered a design for the bridge at Neuville-sur-Ain in the 1760s. Racle is first encountered c.1758 as the protégé and friend of Voltaire (1694–1778), of whose château at Ferney (in the Pays de Gex, Jura) he was the architect. Thanks to their twenty-year connection a good deal is known about Racle, and a substantial body of his papers has been preserved.[31–4]

Voltaire regarded Racle highly and fostered his talents. His principal occupation at this period was an artificial-stone and ceramics factory at Grand Sacconex (then in France, now part of the Swiss Canton of Geneva), where he made large luxurious objects (among them, a sarcophagus intended to contain Voltaire's heart) which Voltaire persuaded his aristocratic friends to buy and employed in his own château. Probably Voltaire had a financial stake in this business, as he also set up a horological colony at Ferney, with artisans from Geneva, specializing in luxurious clocks and watches for the nobility.

Racle's initiation in large-scale civil engineering came in 1768 when, as part of an ambitious property development scheme at Versoix (a village on the then French lake-shore of Geneva), about 28 miles of the little river Versoix were to be canalized in order to make an industrial waterway and harbour, turning Versoix (it was hoped) into a considerable port and city. Voltaire was a prime mover on the spot, while the support of Louis XV was obtained through his chief Minister, the Duc de Choiseul. Racle was appointed engineer and contractor. The canal and harbour walls had been built, and some roads for the new town laid out, when the fall of Choiseul from power in December 1770 caused the collapse of the whole scheme, leaving Racle with obligations he estimated at over 200,000 *livres*.

Over the next few years and through several visits to Paris Racle attempted to clear up the mess, and small sums were obtained for creditors. Eventually a new Minister, Jacques Necker (1732–1804), a former Genevan banker, who corresponded with Voltaire (for whom Paris was still out of bounds), arranged for an investigation to be made by officials of the Ponts et Chaussées. Their report was submitted to Louis XVI (who had succeeded his grandfather in 1774) and after Louis's own visit to Versoix in June 1777 to see that there was no point in continuing the project, a final settlement of nearly 104,000 *lives* was paid on 3 May 1778; the official report had reckoned the total outlay at 800,240 *livres*.[35]

Unable despite protests to recoup his losses in full, and deprived by death of the support of Voltaire, Racle was now (1778) compelled to make a fresh start. If it is true that Voltaire's pen-friend, Catherine the Great, offered him employment in Russia, he refused it. Instead, he closed with the proposal of one Bertin, seigneur of Pont de Vaux, a town on the other side of the Jura, that he should canalize part of the small river Reyssouze connecting the city of Bourg with Mâcon on the Saône and flowing through Pont de Vaux. By the end of 1782 work had begun on a 4 km canal, with a lock at each end, to a plan 'passed through the sieve of' the Ponts et Chaussées in October.[36]

It was at this time that, according to an undated paper by Racle himself apparently written in the winter of 1783/4, one M. Casotte (a friend of M. Bertin) suggested that one of the new bridges crossing the canal might be made of iron.[37] After a few hours' thought Racle had produced a scheme for an iron arch, of which he had a wooden model made. Contrary to later rumours that he had taken his design from that of Vincent de Montpetit, he admitted that he was aware of it but said he knew nothing of its details, until August 1783 when he was shown Montpetit's *Prospectus*. He found that their ideas were completely different, and claimed that M. Gauthey, Chief Engineer of the Ponts et Chaussées in Burgundy, agreed. Nothing was done to realize Racle's iron bridge at this time; a letter to Perronet (1783) discloses that it would have been of 20′ span and 12′ wide: elsewhere Racle gives the span as 22′.

Other proposals for iron bridges can be sketchily traced through Racle's manuscripts: generally the chronology is uncertain and drawings are absent.

(a)   PONT ST JEAN OVER THE SAÔNE AT LYON: 1783

The collapse of one span in the Pont St Jean (or Pont de l'Archevêché) at Lyon, on the site of the present Pont Tilsitt, was to be rectified in 1783 by building a 40′ span in timber. In February of that year Racle wrote a 'Memoir on iron bridges which it would be of value to the city of Lyon to adopt',[38] the content of which suggests that he had studied Guyton de Morveau's 1779 article in the *Journal de Littérature*.[18] Specifically, he now proposed for the repair of the Pont St Jean an extremely flat iron arch, about 42′ long and 42′ wide, with eight ribs at six-foot centres. Each rib was to consist of rectangular framed voussoirs, 6′ long and 1′ deep, strengthened seemingly by an intermediate vertical bar (Figure 5). The voussoirs were to be bolted together through adjacent upright radii (clefs), while transversely ties located at each voussoir junction would hold the ribs together. A timber deck would be laid directly upon them. In the outer ribs the upright radii would be extended to form posts for the railings, and

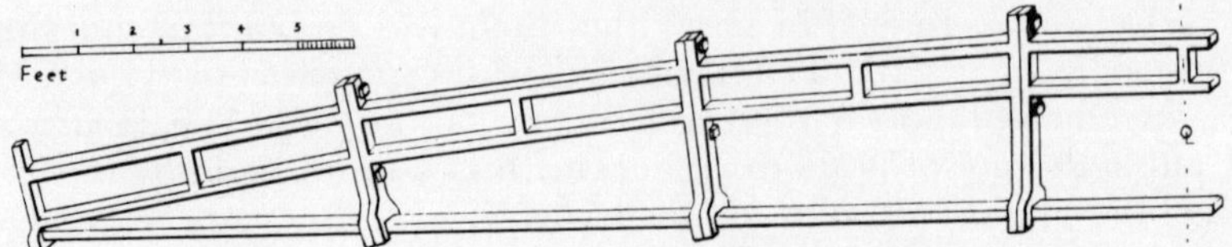

Figure 5. Racle's first system, 1783: early design for a 40-foot span (based on Racle's description).

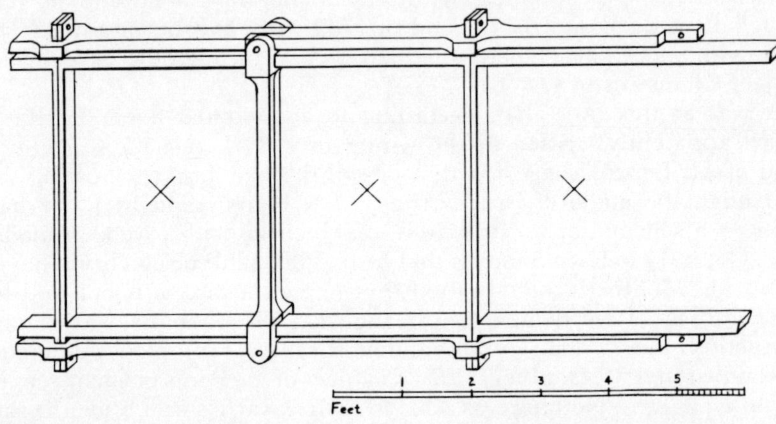

Figure 6. Racle's second system, 1783: detail of 126-foot span design (based on Racle's description and drawings). The crosses indicate the diagonal bracing sketched by Racle, details of which were inadequately worked out in his original drawing.

extensions of them downwards would grip tie-bars intended to absorb the arch thrust. Although Racle does not refer to Calippe it is likely that this idea was derived from him. This iron 'platte-bande' was to be made entirely of 2″ wrought-iron bar, whose weight (bolts and all) Racle put at 39,100 pounds, costing 19,550 *livres*.

At the same time, Racle argued that such short spans constituted a danger to shipping. 'With our iron construction we could eliminate half the piers, providing 80′ spans flanking a central one of 120′, with rises of only 4 or 5 feet.' In consequence, Racle was asked by the Administrator of Lyon, M. Flesselles, to submit a proper design for the complete bridge, which in due course was the subject of a critique by Perronet dated 3 April 1783.[39] This gives the spans as 88 + 147 + 88 feet. Perronet's opinion was that the rise and also the voussoirs themselves were too shallow for the span: he did not think, as Racle did, that the arch-thrust would be wholly taken up by the ties and reckoned that, with unequal arches springing from different heights at the piers, there would be danger of collapse should a tie break. He did not recommend Racle's design, even with a greatly increased rise to the arch or for short spans, but admired the skill of his drawings!

Racle's reply to Perronet survives.[40] It points out that his design had been conditioned by the economical re-use of existing piers, hinted at a new method of fastening the voussoirs together in order to do away with tie-bars, and asked for an opinion of the 20' canal bridge.

(b)   PONT DE LA MULATIÈRE OVER THE SAÔNE AT LYON: 1783

In the same memoir Racle spoke of another failed bridge, the Pont de la Mulatière, whose cost he put at 400,000 *livres*. Reckoning the expense of his own 40' iron bridge at 500 *livres* per linear foot, excluding the masonry, he argued that three of his 'platte-bandes', totalling 240' for the Pont de la Mulatière would cost 120,000 *livres*, plus a similar sum for abutments and piers. Comparing this design with Perronet's masonry bridge at Neuilly-sur-Seine, five 120' spans with 5' rise (1768–74),[41] he wrote that his bridge could be 'infinitely flatter' twice the span and well under half the cost'. Simple foundations and iron columns only 3" in diameter might well serve as piers, since the tied iron arch exerted no thrust.

(c)   PLANS FOR A BRIDGE OF 126' SPAN: PROBABLY LATE 1783

An untitled fragment among Racle's papers with six associated drawings describes a bridge of 126' span, and comments on the designs of Montpetit.[42] Since the bridge has chord tie-bars, its design may be linked with those discussed already, while its greater rib-depth and rise suggest a response to Perronet's comments.

On Montpetit, Racle repudiates the accusation of plagiarism and records his awareness of the misgivings inevitably aroused by Montpetit's long spans: for his part, he thought 100–150 feet ample for navigation, which also helped to make the bridge flatter.

The 126' bridge was to be 42' wide, with 15 ribs 3' apart. Each rib was to be composed of 21 iron voussoirs (6' × 4' deep) forming an arc of 28–30°, that is, with a rise of about 8'. Assembly details of the voussoirs are shown in a perspective sketch, while the other five drawings show a single voussoir and its component parts, illustrating the facility of the method (Figure 6).

The rectangular framing of the voussoirs was made up of 2" square longitudinal members and 2" × 1" uprights (coupled pairs forming radii extended upwards and downwards exactly as in the 40' bridge); further strength was provided by a 2" square central upright (*contraclef*) and 2" × 1" diagonal bracing of the two smaller rectangles so formed. Racle did not show clearly how the ends of the diagonals were to be fixed into the corners.

Laterally, 2" × 1" crossbars (*traversiers*) at top and bottom of the voussoirs connected the ribs; these were in turn stabilized by further diagonal bracing both horizontal and vertical so that the framework was trussed in all three planes by what Racle called St Andrew's crosses. This bracing prevented the possibility of a pedestrian lower deck resting on the bottoms of the voussoirs, as proposed by Montpetit, but Racle pointed out that such a deck merely encouraged vandals to tamper with the structure. The actual road deck, to consist of two courses of timber and a layer of roadstone, was to rest directly on top of the ribs.

Racle still employed a few bolts in this bridge but to lock the voussoirs together without bolts penetrating the radii he relied on 'sleeve-ties' (*tirans-à-bague*), 2″ square bars fitted at the intrados and extrados of each rib, overlapping the joints between the pairs of voussoirs and spanning from one medial upright to the next. At their ends they were grasped by claws at the ends of the medial uprights while, at the centres, they have sleeves or collars fitting over stub extensions of the radii. These sleeve-ties were regarded by Racle as the chief part of his invention, and as making the voussoirs absolutely inseparable. In later designs he omitted the full-length chord tie-bars, mistakenly supposing that the sleeve-ties would bind the bridge into a single rigid frame, which would still exert no thrust on its abutments. The bridge was to be assembled on wooden falsework, supported by piles, as usual.

Racle calculated the weight of iron at 482,520 pounds and quoted the price of wrought iron at 10 *sols* per pound, remarking that if cast iron were to be used the cost could be halved, or the mass of iron doubled.

(d)   THE TOULOUSE BRIDGE COMPETITION: 1785/6

In the autumn of 1783 the Académie des Sciences of Toulouse proposed as the subject of their 1785 prize competition: 'To determine the means of constructing a wooden bridge with a 24ft roadway and a single span over a river 450ft wide the banks of which are 25ft above the water level'.[43] It was specified that the 'oak or pine employed in the structure must not be longer than 30ft nor thicker than 15 inches'. The bridge was to support two moving vehicles each with a load of 6,000 pounds.

Unfortunately the results of this competition, in the event postponed until 1786, were not officially recorded nor is it mentioned in the Académie's *Mémoires*.

According to contemporary press reports,[44] the announced prize was won by 'M. Aubry, Chief Engineer of the Provinces of Bresse and Bugey' (see section 8 below); a 'reserve prize' was divided between competitors numbered 4 and 12, and an honourable mention went to M. Migneron de Brocqueville. The identity of No. 4 is unknown, but No. 12 was Racle. According to Racle's biographer, Jarrin, the reserve prize came from a waterworks fund, and he took this to be the principal prize; he made a second mistake in supposing Racle to have received the prize for an iron bridge design. From references in Racle's own papers, however, it is certain that he received the prize for a wooden bridge (in accordance with the terms of the competition), and a drawing for this is extant though the

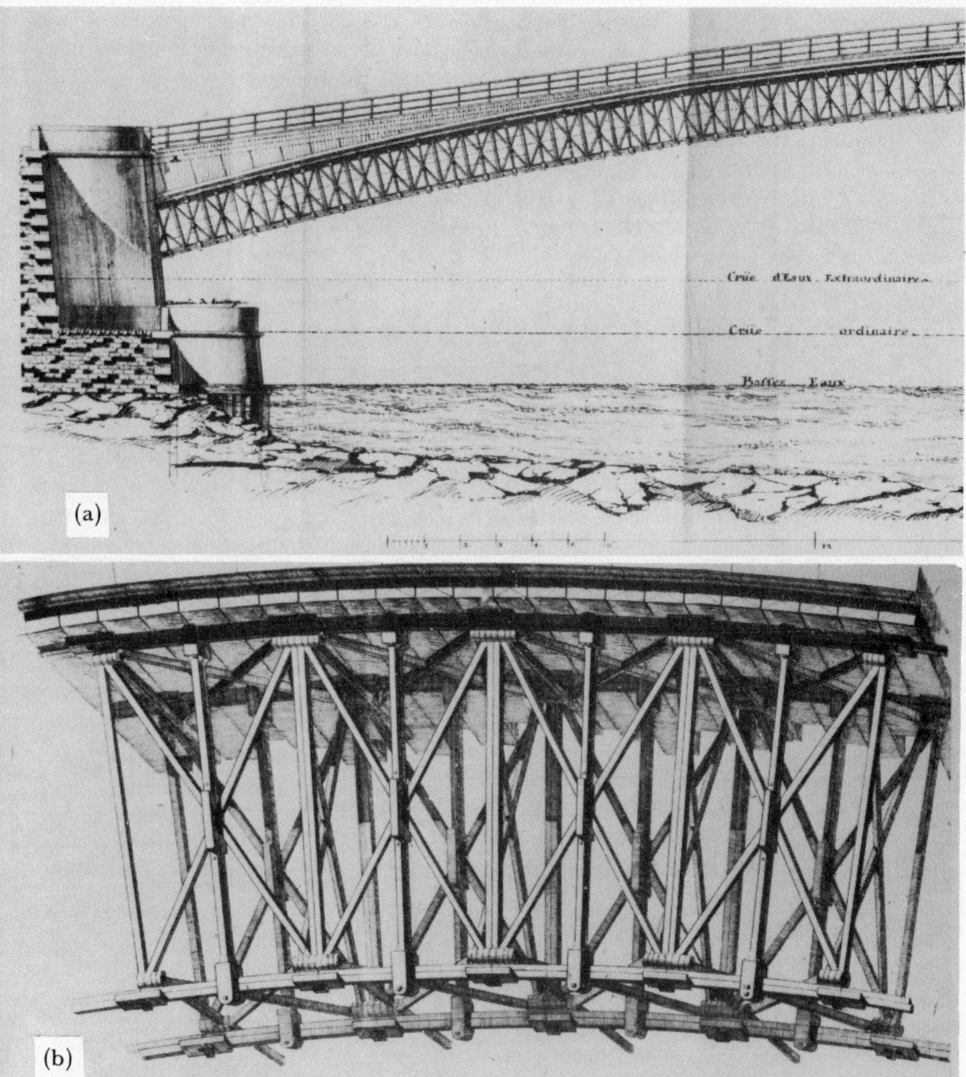

Figure 7. Racle's drawings of his 400-foot bridge for the Toulouse competition, 1785/6: (a) half span elevation, (b) detail.

accompanying memoir submitted to the Académie is lost; he did at the same time offer an unsolicited design for an iron bridge of the same character.[45] Each rib in the timber span of 450′ has a rise of 37½′ and is composed of 30 voussoirs, each 15′ long, diagonally braced to form a single St Andrew's cross.[46] Either from consideration of the lesser stresses at the centre or (more probably) in order to diminish the rise, these voussoirs are reduced in depth to 12′ at the crown from 24′ at the sides; they are held together at top and bottom by overlapping longitudinal members corresponding to the sleeve-ties of Racle's iron bridges.

In the related 400′ iron bridge design (clearly developed from the 126′ span just described, but with the chord ties omitted) the rise was less (30′) and the voussoirs smaller (5′ long × 8′ deep) and thus more numerous (82);[47] they were apparently to be assembled from 2″ square iron. The voussoirs were again diagonally braced with central uprights, joined by sleeve- ties at top and bottom. Towards the haunches of the bridge the radii are extended upwards into sketchily indicated spandrels evidently intended to reduce the deck gradient (Figure 7).

According to one report,[48] after his success in this competition Racle wrote to Catherine the Great offering to design a single arch to span the Neva at St Petersburg, but no detailed plans survive.

(e)   PONT DE CHAZEY SINGLE OR TRIPLE SPAN: PROBABLY c. 1787

The Pont de Chazey over the Ain consisted of two decrepit timber structures, one with a total length of 280′, which the authorities already planned to replace. Knowing this, Racle addressed to M. Feydeau de Brou, Administrator of Burgundy, an obsequiously worded memorial apologizing for his temerity in offering an unsolicited design, justified only by its novelty, and asking for expert criticism.[49] He started from the belief that the existing bridge abutments were too weak to carry a masonry bridge: he therefore proposed the construction of two new piers in the river, the three gaps to be bridged by iron arches each of about 88′ span with a 5′ rise. The spans were to be formed of nine ribs, placed 3′ apart to give a 24′ roadway, each rib being made of 15 voussoirs (6′ × 3′ deep) in the form of St Andrew's crosses, as before, held together by sleeve-ties. Oak beams would carry the road material. No drawing survives.

However, in a different paper or 'Summary memoir on a new system of iron or wooden bridge arches of 280′ span . . . which could be extended to 500′ span',[50] this time addressed to Mgr. Amelot de Chaillon, the provincial administrator, Racle offered to construct the bridge as a single span in either material. No details or drawings are recorded. Making rather different calculations — for he takes the price of iron variously at 48, 55, and 56 *livres* per 100 pounds — the single timber span was the cheapest at a maximum of 150,000 *livres*, the triple iron span next at 158,460 *livres* (including 30,000 *livres* for the two piers) and the single iron span most costly at 240,000–280,000 *livres*. This second paper is different from the first in that, omitting technical details, it contains a discussion of the merits of iron as a bridge material. It gives Racle's answer to ten objections which he said were raised by the prejudiced and uninformed, viz:

1. The brittleness of iron and risk of fracture under load and during frost

2. Rust.
3. Expansion and contraction with temperature change.
4. Compressive stresses in large bridge arches.
5. Failure due to bolt breakages.
6. Weight.
7. Cost.
8. Life.
9. Elasticity and susceptibility to vibrations from vehicles.
10. Vulnerability of bolted fastenings to vandalism.

Much of Racle's reasoning is unconvincing. Most of his faith in his arches depended on his belief that iron was virtually incompressible and he seems to have been totally unaware that buckling was a problem — particularly with his slender bars. He also continued to assert that his arches would have no thrust on the abutments, attempting to justify this statement by reference to his small model which appeared to be free of thrust and stood unbuttressed.

One interesting argument in favour of iron which he advanced was its recyclability. After noting that a common objection against the greater use of iron was the destruction of forests during smelting, he claimed that this was a short-sighted view. A timber roof lasted only about 80 years and then new wood was needed to replace it: if the wood was used to make iron instead, then the resultant roof would last for ten centuries and even then it could be reworked.

In fact, the Pont de Chazey was rebuilt by the official engineer.[51]

(f) PONT DES CORDELIERS AT PONT DE VAUX: 1789

While Racle styled himself 'architect and engineer to the Pont de Vaux navigation canal' his main activity was still in the ceramics factory, which he had moved to Pont de Vaux from Grand Sacconex in 1785 (says Jarrin); in 1789 the province granted him a loan of 6,000 *livres* for six years without interest because he employed 22 people in the works. On 16 January in that same year the wooden bridge over the Reyssouze in the town, weakened by rot for some time although it had been built by Aubry only in 1768, was demolished by ice floes. A week later Racle composed his first memoir on its replacement,[52] no doubt feeling that his chance had come at last, so near home, with Bertin and other dignitaries supporting him. The town (according to Racle) would have to provide three-quarters of the cost of the bridge, the province contributing the balance; his case was that a stone bridge would be exceedingly expensive (62,000 *livres*) while another short-lived timber structure would be absurd. He himself, for 24,000 *livres*, would furnish on the existing abutments a 66′ arch with seven ribs, to carry a 24′ roadway, all to be constructed as in the 500′ [*sic*] wooden bridge awarded the Toulouse prize. The drawing is, as usual, missing.

The town officials and the provincial engineer, debating the choice between wood and stone, passed Racle's memoir to Philippe Vallée (1746–1825), chief engineer of the province of Bresse et Dombes, whose report (dated 29 March 1789) is now missing, but its hostility to Racle is evident from the latter's reply and also from a further report by Perronet. Evidently Vallée ridiculed Racle's belief that his iron fabrication without chord ties would be free from side-thrust and, having calculated what he thought that thrust would be, denied that the existing abutments could support it. He regarded Racle's sleeve-ties as useless. He almost doubled Racle's cost and he stated the undesirability of using iron at all where stone was plentiful.

In rejoinder, Racle accused Vallée of prejudice in favour of conventional, local materials, quoted his canal experience and Perronet's commendation (of his drawings!) and reaffirmed the stability of his iron arch, witness his wooden model of the Pont de Vaux canal bridge.[53] Then he went on to re-examine the question of costs, introducing, however, quite a new design. He reduced the width of the bridge to 15′ and the number of ribs to six, each of 16 voussoirs, and these were to be of cast, not wrought iron. Here he had profited from discussion with the founders at Le Creusot, in relation to the abortive canal bridge. They were now offering to produce the St Andrew's cross bracing in cast iron as well as the frame of the voussoir, and also cast-iron plates for the deck. He therefore provided three estimates:

1. Bridge with cast frames, wrought-iron bracing, wooden deck:
                              88,850 pounds of iron, cost 26,512 *livres*
2. Bridge with cast frames and bracing, wooden deck:
                              93,595 pounds of iron, cost 21,724 *livres*
3. Bridge with cast frames, bracing and deck plates $1\frac{1}{4}''$ thick
                              153,033 pounds of iron, cost 30,050 *livres*

The prices of wrought-iron he here took as 45 *livres* per 100 pounds and cast iron as 18 *livres*: Vallée had assumed 40 and 30 *livres* respectively. The table of weights of the various parts indicates that Racle meant to reduce the depth of his voussoirs from about 8′ at the springings to about 2′ at the crown, and to make his sleeve-ties of cast iron.

The central administration of the Ponts et Chaussées was now consulted because the provincial administrator required, in view of Vallée's damning report, that its approval of Racle's scheme be obtained before giving any financial support, should the town wish to back Racle further. Perronet's reply of 10 August 1789 fully endorsed Vallée.[54] Repeating Vallée's thrust calculations by two different methods he found Vallée's estimate rather too small, and the abutments too weak for Racle's bridge by a factor of two. He agreed with Vallée's criticism of the sleeve-tie bars: it would be better to thicken the main frames, if necessary, rather than add a parallel set of bars. Nevertheless he felt that the sleeves themselves were useful to keep the voussoirs together. He recommended making and testing a full-scale voussoir assembly. Perronet again advised widening the bridge towards the haunches to prevent warping, 'as proposed by Parent in 1699 for the wooden Sèvres bridge' and also the addition of ties at the top of the springing joints to restrain the ironwork from tending to part from the abutments. His summary conclusions were as follows:

> From the above reflections we are of the opinion that,
> 1) It would be possible to erect an iron arch at Pont de Vaux, and that M. Racle's system is ingenious and merits congratulations;
> 2) The existing piers of the wooden bridge will need to be at least doubled in thickness to sustain M. Racle's bridge;
> 3) Practical tests are required first to study the compressive stresses on the voussoirs and the strength necessary in the tenons and collars.
> 4) The sleeve-ties and cross-ties should be omitted;
> 5) The arch should be laterally buttressed by adopting M. Parent's system;
> 6) Iron ties should be placed at the top of the springings to resist parting between the iron and the masonry.

In a final sentence he added, 'We conclude by observing that it would be simpler and cheaper to construct a wooden bridge of two or three spans with intermediate piers than a single iron span of 66'. This was the deathblow to Racle's plan and it appears that another timber bridge was erected at Pont de Vaux.

Hence Racle's biographers, Jarrin and Amanton, were both mistaken (despite the documents they cite) in stating that an iron river-bridge for Pont de Vaux was fabricated in Racle's yard. The only bridge erected was the small one for the canal, recorded by neither biographer.

(g)   BRIDGE FOR THE CANAL DE PONT DE VAUX: 1790

Work on 'M. Bertin's Canal' went slowly after the enthusiastic start in 1782, so that the iron bridge which had begun Racle's series of designs was postponed for several years. Probably it was in the end undertaken to justify Racle's confidence and to provide a visual example, necessary to win contracts for bigger bridges.

The span of this bridge was about 20', although the canal itself (according to Depery) was 14m wide; hence the crossing must presumably have been intended for one of the locks. Although Racle had originally proposed to build in wrought iron, by the end of 1788 his thoughts had turned to the use of cast iron instead, no doubt directed (as noted above) by the iron-founders at Le Creusot to whom he had gone for estimates. These Fonderies Royales de Montcenis resulted from proposals made by William Wilkinson to Louis XVI in 1781/2, though the direction of the works fell to his collaborators Toufaire and de Wendel. The first of the English-style coke furnaces there was blown in 1785, and cannon-founding begun in 1788, so that Le Creusot was a very recent establishment when Racle turned to it.[55] By January 1789 (according to one of his own undated drafts) Racle was able to show the engineer Vallée a full-sized wooden assembly of one rib of the canal bridge, the pieces ($1\frac{1}{2}''$ square the section) being intended as patterns for the founders' moulds. Racle gave it as his intention to subject the wooden model, after its return from the Montcenis foundry, to loading tests before the Société d'Emulation de Bourg.[56] He added that Wendel, 'Director-General' of Le Creusot, who had seen the Coalbrookdale arch, had welcomed his own new system of construction; while in another paper he attributes to the Le Creusot founders the suggestion that cast iron be used entirely for the voussoirs, as already mentioned. Amanton referred to an agreement for casting bridge parts signed at Le Creusot on 25 April 1789; this clearly related to the canal bridge, not the abortive river-bridge at Pont de Vaux as he thought. This is confirmed by Jarrin's discovery of the delivery note relating to these castings, headed 'Chalon-sur-Saône 31 October 1790' where their weight is given as 18,621 pounds 'du marc', sufficient only for a very small bridge.

The bridge was never placed over the canal. Racle died on 8 January 1791, and Bertin in the next year. Amid the confusion and mob-violence of the revolutionary years it vanished. Amanton in 1808 was told at Pont de Vaux that some time after the bridge had been set up in Racle's yard, 'it was destroyed and broken up during the night in such a manner that the authorities had no chance to interfere'. The ruins were then sold for scrap by Bertin's successor as landed proprietor at Pont de Vaux; this would put the destruction, probably, in the violent year 1793. At that time the Frerejean brothers had recently established a foundry and bell-metal refinery in the town, at which a steam-engine was installed in 1796, which might have welcomed such iron scrap.

Although other accounts[57] refer to the existence of the canal bridge in 1796 or later, these turn out on analysis to be founded upon mistakes and may safely be dismissed. Depery, writing in 1835, said that the canal was still unfinished although the lower lock was in service.

(h) MISCELLANEOUS DESIGNS

A drawing entitled 'Elevation of four cast-iron voussoirs for bridge arches of any desired span', probably belonging to the Pont de Vaux period of 1789, exists among Racle's papers (Figure 8).[58] This is clearly not the canal bridge; it shows voussoirs twice as deep as wide, with diagonal as well as upright members and the usual sleeve-ties. There are also undated drawings for roofs of wood, wrought iron and cast iron 'for any desired span', all based on the tied-arch principle with wrought-iron chord bars (Figure 9).[59] Other roof designs, intended for dry docks, were submitted to the naval authorities by Racle in 1789/90 though produced originally for a competition in 1786.[60]

The strangest of Racle's drawings, together with lengthy drafts and

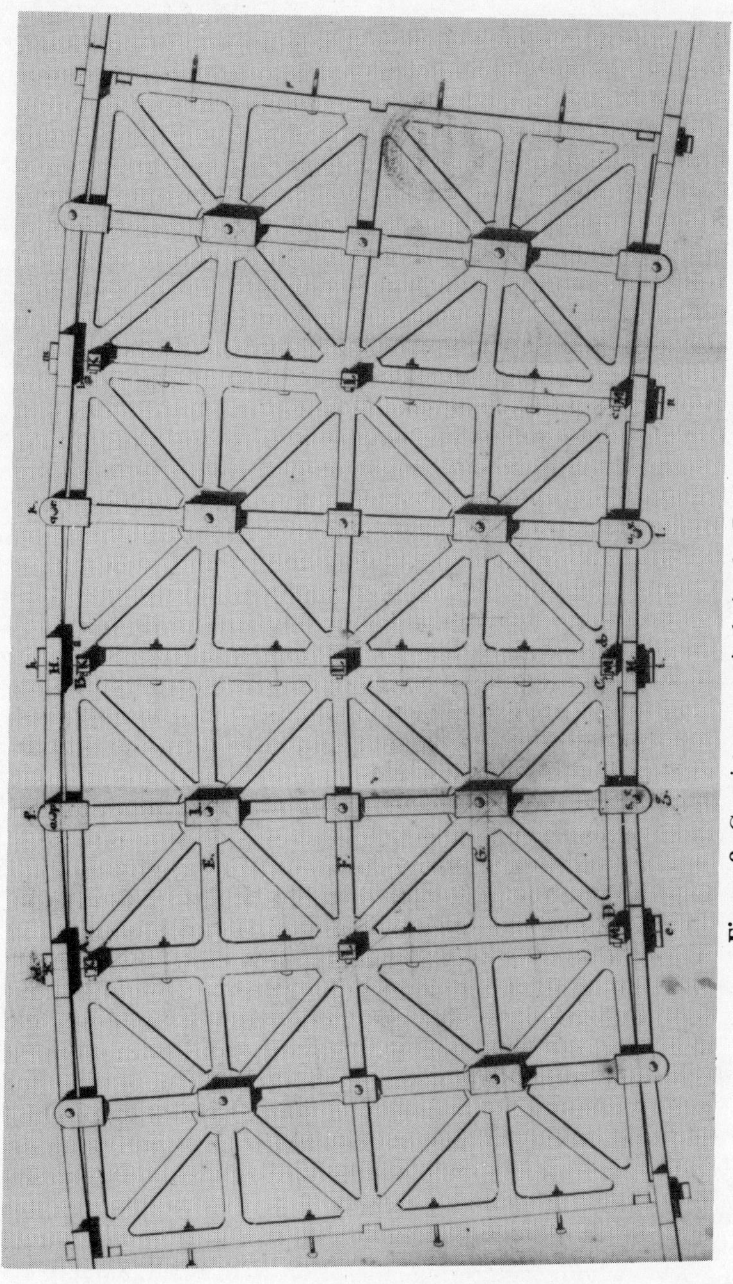

Figure 8. Cast-iron voussoir design by Racle, *c.*1789.

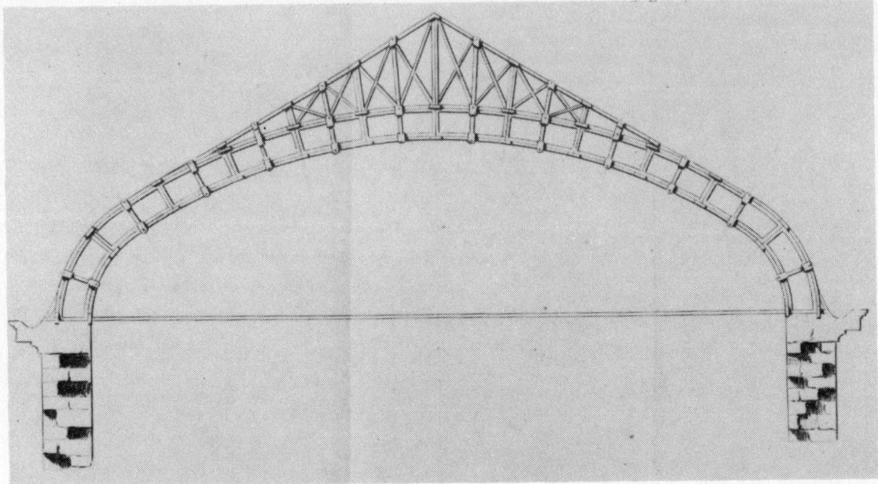

Figure 9. Iron roof design by Racle, *c*. 1785–9.

correspondence, relate to incomprehensible proposals to apply epicycloidal gearing to what looks suspiciously like a perpetual motion machine.[61] It is possible that mental instability, disappointment, financial worries and revolutionary disturbance all contributed to bring about his death at a fairly early age.

8.   Proposals by Nicolas Aubry: 1780s

Among the would-be constructors of iron bridges before the Revolution the only professional and qualified engineer was Nicolas Aubry (1727–99). He was a highly competent member of the Ponts et Chaussées whose important papers were well known to his successors in the next century, though today he is a shadowy figure, his reports buried (largely in the archives of the Département de l'Ain) and his publications hard to find. The most recent French biographical summary,[62] following Dufay,[63] gives Bourg as the place of his birth, about 1747. This date is obviously wrong; the records of the Ponts et Chaussées give his birth year as 1724 or 1725. A manuscript amendment in the Ain archive copy of Dufay states that he was born at Dormans (Soissons diocese) on 3 January 1727 and that he was married at Bourg on 20 November 1770.[64] From the Ponts et Chaussées records Dartein reported that he joined that service in 1750, became assistant engineer at Grenoble in 1754, and in 1762 submitted two well-conceived but unrealized designs for arches over the Durance, at Embrun and near Savines, each having a span of 120′. In 1766 he became engineer of Bresse et Bugey, a post he occupied for over 20-years. Since Racle — whose engineering enterprises were mostly undertaken in Aubry's shadow — credits him with the design of the wooden Pont de

Vaux bridge (66′ span, 1768) this must have been one of his earliest works in the province.

Throughout the 1760s he regularly attended the meetings of the Académie Royale d'Architecture at Paris, and he was sometimes joined with Perronet on special committees. When construction of his most famous bridge, at Neuville-sur-Ain, began in 1770 Aubry's attendance at the Académie ceased. This bridge, illustrated and described in detail by Dartein,[65] took five years and 300,000 *livres* to build. It has two shallow masonry arches of 29.3m span and a comparatively slender central pier. In 1773 Vallée, who had worked on the Neuilly bridge with Perronet, came to work on the Neuville bridge as assistant engineer in Bresse et Bugey. This was also the year in which Aubry designed the Bellegarde bridge over the Valserine near its confluence with the Rhône, an arch of 66′ span and 70′ high.

Aubry first encountered Racle over the development of Versoix (see previous section). He was responsible for a survey of the canal in 1773, and perhaps for the survey (engraved in 1774) of the new town, which he was certainly instructed to mark out before the visit of Louis XVI in June 1777. As Chief Engineer of the Province Aubry also effected the final winding-up of the scheme for Necker (1777/8).

As already noted in section 7 (d) above, Aubry won first prize in the Toulouse bridge competition (1785/6), but his published description of his design, like his papers on the adhesion of mortar and the statics of arches, has proved untraceable.[66,67] Like Racle and Guyton de Morveau, Aubry was a member of the Société d'Emulation at Bourg, from whom in January 1787 he received a prize for a memoir on prevention of flooding in the Reyssouze river.[68] It was somewhat after this that, according to Dufay, Aubry was recalled to Paris to become an Inspector of the Ponts et Chaussées, while Dartein describes him more specifically as 'inspecteur général des turcies et levées' for about two years from 1789.[69] During these last years of work Aubry published the untraced Toulouse bridge paper and a large *Memoir(s) on various questions relating to the science of public works and economics*, quoted by Rondelet, Gauthey and Navier for its data on the strengths of timber and iron, but also not to be found in the libraries explored.[70]

Since it contains Aubry's iron bridge proposals the loss of this book is highly unfortunate, since without it information can only be taken from the unreliable Gauthey–Navier discussion of 1813. This relates that 'A bridge of six arches of 23m [i.e. 70 ft] span, proposed by M. Aubry for Cordon in old Bresse, was to have been constructed of wrought iron. The ribs were to be formed by two arcs of different radii, connected by *moises pendants* extended and braced to carry the deck'. We may read Cerdon, a town a few miles from Neuville, for Cordon and put the date at 1785–7.[71]

Again according to Gauthey, in the Toulouse competition Aubry put forward an additional 'plan for the construction of a 97m [i.e. 300 ft] span wrought-iron arch. The ribs were to be formed of two arcs connected by *montants normaux* with diagonally-braced panels. To prevent horizontal movement M. Aubry, in addition to various braces, employs what he calls *pendentives*, that is to say, arched pieces of wrought iron which connect the deck and the side walls and which are held in the horizontal plane of the deck by brackets . . .'. Presumably none of these plans for iron bridges came to fruition. Aubry retired just before the Revolution and died at Fontainebleau in 1799.

9. Bélanger

Some French architectural historians have stated that the architect François Joseph Bélanger (1744–1818) designed an iron bridge for the Seine in the 1780s. Though the plan is not improbable no evidence confirming it has been found. Bélanger, at this period architect to the Comte d'Artois ('chef des Anglomanes'), was certainly interested in iron construction. He designed the buildings for the two waterworks at Paris in which the Wilkinsons were concerned with the Périers during 1777–86.[72] In conjunction with a *serrurier* named Deumier he proposed, in 1782, a roof for the Paris corn-market in the form of a dome of copper on wrought-iron ribs; the wooden dome using ribs of de l'Orme's type that was actually built was burnt out in 1802. Bélanger again proposed an iron-ribbed dome in that year which after much debate and modification was approved in 1807 and fabricated at Le Creusot in 1809–11.

Bélanger was also the designer of several small wooden footbridges in the Palladian and Chinese styles for the gardens of Beaumarchais, Sainte-James and at Bagatelle in the 1780s. In 1787 he became architect to a company headed by his friend and patron Beaumarchais, which had received a concession to build a major new bridge over the Seine near the Jardin du Roi (today the site of the Pont d'Austerlitz). Plans for such a bridge had been formed as far back as 1773, when Perronet had unsuccessfully proposed a bridge with seven 90 ft wooden spans on piles. Coincidentally, it was in the summer of 1787 that Thomas Paine arrived in Paris with his model of an iron bridge,[73] and thanks to Franklin's recommendation persuaded Jean-Baptiste Le Roy to present it to the Académie Royale d'Architecture in July.[74] Le Roy arranged for a sympathetic committee to report upon it, rather than it have it scrutinized by the more critical Ponts et Chaussées experts led by Perronet.[75] Its views were produced on 29 August and Paine then left for England to promote his bridge there,[76] leaving his French friends to try and supplant Bélanger's design by Paine's.

Bélanger's design, we learn from an article (January 1788) by a M. Charon [a pseudonym?] commenting on the rival bridges, was for five timber arches of 120 ft span, with a very flat roadway for vehicles and covered galleries for the use of pedestrians and flower-sellers, to be built upon cast-iron columns set in the river bed.[77] This plan M. Charon vastly preferred, since all who had seen the Coalbrookdale design had been

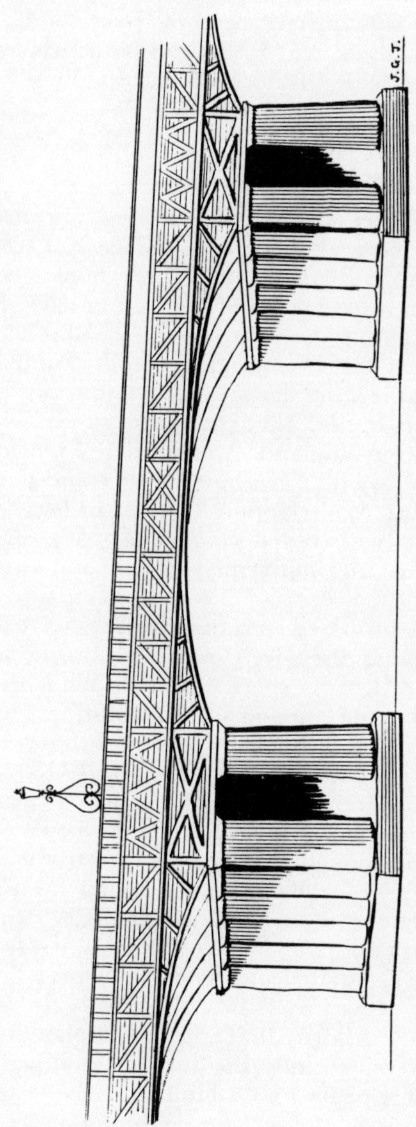

Figure 10. Bélanger's design for a Seine bridge, 1787 (based on the watercolour sketch by Maréchal).

aghast at its frailty; how much more alarming, therefore, an iron skeleton six times the length of that arch would be!

Although it has been said[78] that Bélanger proposed a bridge wholly of iron for this site, it is clear from Maréchal's water-colour sketch that the design superstructure was to have been of wood, as M. Charon wrote. Figure 10 is based on this sketch. In the end both Paine and Bélanger as well as other aspirants were displaced,[79] and the final form of the Pont d'Austerlitz was not evolved until after 1800.

10. Conclusions

For lack of information about the designs for iron bridges in pre-revolutionary France, and adequate evidence of success with the new material in that country, British writers have tended to disparage the French contribution to the early history of iron bridges. However, it will now be clear that in the interval between the Coalbrookdale Bridge and the general revival of interest in iron bridges in Britain that followed from the exhibition of Paine's model at Paddington in 1790/1, several important ideas were under discussion in France which were later developed with success in England. Among these French design-features were iron voussoir arches with or without mortice-and-tenon interlocking, iron voussoirs with overlapping wrought-iron straps, deep double-membered ribs with radial connecting bars, single and double diagonal panel-bracing, iron bowstring girders, and cast-iron columnar piers.

Since most of these proposed structural features were openly discussed and displayed as models they were accessible to foreigners and were undoubtedly known to British experts. M. de la Blancherie of Paris, who first publicized the ideas of Calippe and Montpetit, was well known to the London Society of Arts to which he presented busts of Perronet and Franklin. Paine's British patent for iron bridges of 1788 followed his visit to Paris and the publication of Montpetit's final paper on this theme. Racle's work may well have reached England via Le Creusot and the Wilkinsons.[80] William Reynolds of Ketley, who introduced Telford to the mysteries of structural iron in the Longdon-on-Tern aqueduct (1795) — which is a series of *platte-bands* with auxiliary struts — is known to have been familiar with events in France.[81] Nash's first iron bridge (1795) and his patent (1797) show clear French influences.[82] Rennie and his adviser on bridges, Charles Hutton, were both acquainted with Montpetit's *Prospectus*.[83] Above all, Thomas Wilson's Sunderland bridge (1792–6) obviously stemmed from the ideas of Montpetit.[84]

It is hoped that this paper may stimulate further work in the French archives, and that more illustrations and information may come to light.

Appendix 1: Places and People

The majority of places named, other than those in or around Paris, are in

the old Burgundy region of France. In view of their relative unfamiliarity
a map is given. Modern Burgundy is taken to comprise the *départements* of
Yonne, Côte d'Or and Saône-et-Loire, lying generally west of the Saône.
Before the Revolution, however, Burgundy also extended eastward along
the Rhône as far as the Swiss border, taking in the districts of Bresse,
Dombes, Bugey, and the Pays de Gex, all now parts of the *département* of
Ain.

The most famous of early French structural ironwork pioneers was
Soufflot (1713–80) who was born near Auxerre (Yonne) and came to fame
at Lyon in the mid-eighteenth century. Among those actually born in Lyon
were Soufflot's contemporary Goiffon (1712–76), Soufflot's pupil Rondelet
(1734–1824), Perronet's pupil Gauthey (1732–1806), and Montucla
(1725–99).

Vincent de Montpetit (1713–1800) conceived his first iron bridge plan
with Goiffon at Lyon but was himself born upstream of Mâcon. Aubry
(1727–99) was married at Bourg. Racle (1736–91), Guyton de Morveau
(1737–1816) and Gauthey's nephew Navier (1785–1836) were all born
further north at Dijon. Buffon (1707–88) was born at Montbard.

The ironworks of Le Creusot lies west of the Saône. Voltaire's Ferney
still lies within the French border on the east side but Versoix and Grand
Sacconnex are now Swiss territory.

Appendix 2:   Weights, Measures and Money

Weights, lengths and prices have been deliberately left unconverted from
the old French units although most of the terms have been anglicized, i.e.
'*pied*' translated as 'foot'. For those wishing to convert to British imperial
or metric measures the following brief notes should be adequate.

Weights. The terms *once* and *livre* have been translated as 'ounce' and
'pound' and are very roughly equivalent to the British units. For more
accurate conversion:

> 1 old French pound = 1.086 English pounds (17.375 English ounces)
> = 493 grammes metric

Lengths. The terms *ligne*, *pouce*, *pied* and *toise* have been translated as
'line', 'inch', 'foot' and 'fathom', and are very roughly equivalent to the
British units. For more accurate conversion:

> 1 old French foot = 1.066 English feet (12.792 English inches)
> = 0.325 metres metric

Monetary units. The term *livre* has not been translated as 'pound' partly to
avoid confusion with the pound weight but primarily to emphasize the
disparity between French and English monetary values. When the *livre*
was re-named the franc after the Revolution it was worth about $9\frac{1}{2}$ old
pence, i.e. only about £0.04. The other French units were the louis (= 20
livres), the *écu* (= 3 *livres*) and the *sol* (= 0.05 *livres*).

Notes

Illustrations 1, 2, 3, 5, 6 and 10 were drawn by the author. See note 34 for 7, 8 and 9.

1. *Leeds Intelligencer*, 2 Jan 1770: 'A few days ago was finished by Mr. Tobin of this town, a most curious bridge of one arch 6 feet wide and 72 feet in span; made entirely of iron . . . over a canal in Sir George Armytage's park at Kirklees. It also has iron ballustrades which are ornamented with roses of the same metal . . .'. This bridge was mentioned in Gardner, J.S. *English iron work of the XVIIth and XVIIIth centuries*, 1911, p. 254, but the claim was not taken seriously until J. Goodchild of the Cusworth Hall Museum, Doncaster drew attention to the newspaper confirmation a few years ago and established that Maurice Tobin, a whitesmith and ironmonger, died September 1773 and that his son gave up the business in 1774, (letter from J. Goodchild to R. Paxton).

2. Ferchault de Réaumur, R.A. *L'art d'adoucir le fer fondu*, Paris, 1722; new edition 1762. For English translation see Sisco, A.G. *Réaumur's memoirs on steel and iron*, Chicago, 1956. Réaumur, born in Rochelle the same year as Desaguliers, drew attention to the value of cast iron for repetitious production of parts.

3. Perrault, C. *Recueil de plusieurs machines de nouvelle invention*, Paris, 1700. This was reprinted in Perrault, C. & P. *Oeuvres diverses de physique et de méchanique*, Leyden, 1721, vol. 2, pp. 712–14, Plates 10 and 11; and again in Gallon, G. *Machines et inventions approuvées par l'Académie Royale des Sciences* etc., 1735–77, vol. 1, p. 59, Plate 14.

4. Ozanam, J. *Récréations mathématiques et physiques*, new edition augmented by 'M. de C.G.F.' (pseudonym of Montucla, J.E., since printing approval, dated 1775, was signed by Montucla himself in his role as censor), 1778, vol. 3, pp. 374–8 and Plate 4, Fig. 20. Montucla claimed to have conceived the design in Peru after experiencing a crossing in a *tarabita* (rope and basket). This book went through innumerable editions and an English translation was published by Hutton in 1803.

5. *Procès verbaux de l'Académie Royale d'Architecture*, 1924, vol. 8, pp. 103–7 (22 and 29 July and 2 Sept. 1771). For Grubenmann bridges see Killer, J. *Die Werke der Baumeister Grubenmann*, Zürich, 1941 and 2nd edition 1959.

6. The French claims gave rise to a retaliatory English hoax that iron bridges had been proposed for the Thames in the seventeenth century. This was launched in the *European Magazine*, 1825 (Sept.), 20–7 and reprinted in the *Mirror*, 1825 (Oct), 291–5. This article was a spuriously expanded version of a genuine parliamentary debate reported in Grey, A. *Debates in the House of Commons from the year 1667 to the year 1694*, 1763, vol. 1, pp. 416–7. It was exposed in Thomson, R. *Chronicles of London Bridge*, 1827, pp. 607–8.

7. Gauthey, E.M. *Traité de la construction de ponts*, Paris, 1809–16, 3 vols. ('Des ponts de fer' in vol. 2, 1813, Book 3, p. 113). The reference to Italian works no doubt relates to the posthumous volume of designs by Faustus Verantius (1504–73) which has been variously dated 1590–1625 and was reprinted Munich, 1965. In addition to the well-known chain bridge designs this book had a design for a bell-metal bowstring girder.

8. *Encyclopédie méthodique*, vol. 3 of the 'Architecture' division, 1825, article 'Pont', p. 167.

9. Dartein, F. de, *Études sur les ponts en pierre* etc., Paris, 1907–12, 4 vols, vol. 2, p. 42. I am grateful to P.F.B. Alsop for supplying me with several xeroxed extracts from this scarce work.

10. Mook-Aray, A. *La reconstruction des ponts de Lyon*, Circulaire Série K, No. 13, Inst. Techn. du Bâtiment et des Trav. Publ., Paris 1946.

11. Guillemain, C. Article in *Nouvelle Revue du Lyonnais*, Sept. 1932. I am indebted to Mme A. Blanchet of the Bibliothèques de la Ville de Lyon for sending me a short extract from this and for copies of the articles cited in note 13.

12. Gautier, H. *Traité des ponts*, Paris, 1716 (later editions 1723, 1728. 1755, 1765, 1809), pp. 171–4 and Plate 17. Guillemain stated that the design is 'attributed to one Aubert' locally.

13. 'N', Article dated 7 April in *Bulletin de Lyon*, 1807 (11 April), pp. 114–15. The gist of this was given wider circulation in the *Gazette Nationale*, 1807 (26 April), p. 455.

14. Breghot du Lut, C. *Biographie Lyonnaise: Catalogue des Lyonnais dignes de mémoire* etc., Paris, 1839.

15. Dumas, J.B. *Histoire de l'Académie . . . de Lyon*, 1839.

16. See for instance *Biographie nouvelle des contemporains*, 1824; Michaud's *Biographie universelle*, 1849; and *Nouvelle biographie générale*, 1968.

17. Lemonnier, H. (ed.) *Procès verbaux de l'Académie Royale d'Architecture*, 10 vols. 1911–29, vol. 9, p. 128, entry for 29 March 1784 notes Calippe's presentation of his ideas but gives no details, Gauthey (*op. cit.*, p. 114) briefly describes it as a bow-string bridge.

18. Guyton de Morveau, L.B. 'Lettre de M. de Morveau sur la possibilité et la manière de construire les ponts tout en fer', *Journal de Littérature, des Sciences et des Arts*, 1779, *5*, 278–88. Six volumes appeared each year.

19. Fulton, R. *Treatise on the improvement of canal navigation* etc., 1796, Plate 14, Figs. 1 to 4. Fulton went to France in 1797.

20. A model of Ango's beam was referred to in *Journal de Paris*, 1782, 8 Jan. A full-sized floor was built for reviewing by a group from the Académie Royale d'Architecture in July but, despite tactful wording from the committee, it was evidently unsuccessful: presumably the light ribs were unable to support a conventional heavy floor. When Calippe presented his 600′ span bridge design to the Académie in 1784 it is not surprising that he got a dusty response. However, the subsequent introduction of lightweight hollow-pot floors led to the successful re-promotion of Ango's ribs from 1785 onwards. For illustration of Ango's beam see *Trans. Newcomen Soc.* 1966–7, *39*, Plate 28. I am indebted to Professor A.W. Skempton for lending me his notes on Ango which facilitated research on this point.

21. See note 18. This paper may have been reprinted later because Gauthey cited one which was evidently the same or similar, but gave the reference as 'les Affiches de Bourgogne, 1782'.

22. Vincent de Montpetit, A. 'Réponse à la lettre écrite par M. de Morveau' etc. *J. de Litt. des Sciences et des Arts*, 1779, *6*, 138–44.

23. École des Ponts et Chaussées MS. 2219. This comprises three items, the main one headed, 'Les Commissaires nommés par l'Académie ont examiné les Mémoires et Dessins du Projet proposé par le Sr de Maupetit [*sic*] pour établir la possibilité de la construction d'une arche faite en fer forgé qui auroit au moins 200 pieds d'ouverture et dont le modèle a été mis sous les yeux de l'Académie' (18 pp., no date). Two appendices are 'A. Problème. Le poids d'un des voussoirs de la clef d'une voûte et l'inclination où la coupe de ses joints de lits étant donnés, connoitre le force de la pression qui sera employée de part et d'autre contre ces lits, pour soutenir le voussoir' (10 pp. with diagram, no date); and 'B. Calcul sommaire du prix d'une arche de fer de 200 pieds d'ouverture et 40 pieds de large proposée a l'Académie des Sciences par Mr de Maupetit' (9 pp. no date). The Commissioners are given as Messrs Vaucanson, Montigny, Bezoult, and Perronet but the papers were clearly written by the last-named. They must date from late 1779 or early 1780.

24. Vincent de Montpetit, A. *Prospectus d'un pont de fer d'une seule arche, proposé, depuis vingt toises jusqu'à cent d'ouverture, pour être jeté sur une grande rivière: présenté au Roi le 5 Mai 1783*, Paris, 1783 (publication approval dated 13 May). This was reprinted in vol. 2 of the 'Arts et Métiers Mécaniques' division of the *Encyclopédie Méthodique*, 1783, pp. 638–42 with a plate augmented by some small sketches indicating the assemblage.

25. Vincent de Montpetit, A. 'Observations physico-mécaniques sur la théorie des ponts de fer, d'une seule et grande arche de trois à cinq cens pieds d'ouverture', *J. de Phys.*, 1788, *32* (June), 430–7.

26. See note 2. Vincent said that he had examined some gear wheels with 1″ square teeth made by Réaumur's process.

27. This is clearly one of Montpetit's drawings but it is filed in Ecole des Ponts et Chaussées MS. 2218, which comprises the Racle MSS. It is entitled 'Portion d'arrête d'un pont de fer, sans coupe de voussoir, formé par de simples bandes prolongées et liées ensemble par des espèces de moises' (no signature or date).

28. Perronet had proposed to built two arches of 150′ span each with this stone to cross the Seine at Melun and had already carried out the compression tests.

29. Vincent, Guyton, Perronet and Racle all harped on the problems of thermal stresses and strains and Vincent claimed to have experimented on this subject: he was of course

knowledgable on the behaviour of pendulums in horology. The works of the Leyden professor P. van Musschenbroek dealing with metal-bar pyrometry and the strength of materials had been republished in French some years earlier (*Cours de physique expérimentale et mathématique*, Paris, 1769, vol. 2, pp. 339 *et seq.*, Plate 33). In addition Racle cited an *Essaye sur l'horlogerie* by M. Berthou, 1763.

30. The career of Thomas Paine (1737–1809) in England grew shabbier and shabbier until he emigrated to America in 1774. He emerged from the American revolution (1776–82) as a folk-hero and was given a considerable grant of land and money for his role as chief rabble-rouser. With the wealth and leisure then available he dabbled in bridge design from 1785 onwards. He employed a more recent *émigré* John Hall (an inventive ex-foreman of Boulton & Watt who had been employed at Coalbrookdale during the construction of the Iron Bridge and subsequently at the Walkers' foundry at Rotherham) to make some models: one of wood in the winter of 1785–6, one of cast iron (spring 1786), and one of wrought iron (autumn 1786). He took the last to Paris in the summer of 1787 with influential introductions from Franklin (see also notes 73–5).

31. Most French biographical dictionaries have entries on Racle. In addition to those cited in note 16 see Depery l'Abbé, J.I. *Biographie des hommes célèbres du département de l'Ain*, 1835; and Dufay, C.J. *Dictionnaire biographique des personages notables du département de l'Ain*, Galerie Civile, 1882.

32. Amanton, C.N. *Notice biographique de M. Léonard Racle* (new revised edition), Dijon, 1810.

33. Jarrin, C.H. *Deux oubliés*, Bourg, 1895.

34. I am indebted to the Conservateur of the Bibliothèque Publique de Dijon (P.Gras) who drew my attention to the two large volumes of Racle papers (MSS 442 and 443) in his care and subsequently arranged for me to obtain microfilm copies. They comprise a mass of documents presented by Racle's brother, one-time almoner at Ferney. Although most are unused drafts, many only fragmentary, they have proved invaluable in the present study. There are over a dozen items relating to bridges, cited individually below. The folio numbers were inserted after the papers were bound up and each one covers a pair of facing pages.

35. Copies of letters and accounts survive; Dijon MSS pp. 157–200. Other major documents (not seen) are in the Archives Départmentales de l'Ain at Bourg (Items C1093 and C1094 in index).

36. There are no papers in the Racle MSS dealing directly with the canal but several passing references in others. Major papers exist (not seen) in the Ain archives (Item C1095 in index).

37. Racle, L. Untitled, undated fragment (late 1783 or early 1784): Dijon MSS pp. 236–8.

38. Racle, L. 'Mémoire sur les ponts en fer, qu'il seroit important à la Ville de Lyon d'adopter. Fevr 1783', MS in the Racle dossier (MS 2218) of the Bibliothèque de l'Ecole Nationale des Ponts et Chaussées in Paris.

39. Perronet, J.R. 'Observations faites sur la proposition du Sr Racle de construire un pont de trois arches en fer à Lyon, 3 Avril 1783' Dijon MSS pp. 203–4.

40. Racle, L. 'Réponse aux observations faites par Monsieur Perronet sur la proposition de Racle de construire un pont en trois arches en fer à Lyon, avec priere d'examiner l'application de ce genre de construction aux petits ponts que comporte le canal de navigation que Monsieur Bertin fait éxécuter de Pont de Vaux à la Saône': undated (1783) MS in Ponts et Chaussées dossier (see note 38).

41. Perronet's great book on bridges (*Description des projets et de la construction des ponts de Neuilly, de Mantes, d'Orléans et autres*) had just been published (Paris, 2 vols, 1782–3) and a supplement was issued in 1788. Perronet does not mention iron bridges.

42. Racle, L. Untitled, undated fragment (c. 1783–4): Dijon MSS pp. 230–6. The surviving Racle drawings are all grouped together and numbered consecutively by the collator but Plates I to VI appear to belong to this paper.

43. *J. de Litt. des Sciences et des Arts*, 1783, 5, pp. 69–70.

44. *Journal de Paris*, 1786 (2 Oct.), p. 1135; and *Mercure de France*, 1786 (23 Sept.).

45. There is an introductory fragment entitled 'Mémoire et réflexions sur les différents systèmes des ponts, sur une nouvelle construction métallique applicable aux ponts et bâtiments, et une nouvelle charpente de grande arche en bois' (Dijon MSS pp. 238–40) but this contains merely general discussion.

46. Dijon MSS Plates XII and XIII.

47. Dijon MSS Plates VII and VIII.

48. Anon. 'Extrait d'une lettre d'une habitant de Ferney à M. Racle Architecte-Ingénieur, Genève 18 juillet 1787', *Journal de Paris*, 1787, (9 August), pp. 970–1.

49. Racle, L. 'Mémoire sommaire sur l'application des arches en fer au Pont de Chasey [*sic*] sur la Rivière d'Ain', undated (c. 1786–7): Dijon MSS pp. 222–4.

50. Racle, L. 'Mémoire sommaire sur un nouveau sistème d'arche de pont en fer et en bois de 280 pieds d'ouverture relatif aux desseins ci-joint, et qui peuvent etre portés à 4 ou 500 pieds, applicable aux différentes rivières qui traversent la Bresse et le Bugey', undated (c. 1786–7): Dijon MSS pp. 224–7; and a second version with amended title and text, pp. 228–30.

51. Papers and plans (not seen) relating to the various ponts de Chazey are held in the Ain archives (Item C1087 to C1090 in index). The timber bridge built at this time was illustrated by Gauthey who stated that it was the first ever built with the deck supported on arched ribs made of multiple curved timbers. It had four spans of 60'.

52. Racle, L. 'Mémoire sur le pont des Cordeliers de la Ville de Pont-de-Vaux, Chemin de Mâcon, 23 janvier 1789': Dijon MSS, pp. 204–9.

53. Racle, L. 'Réponse de Racle au rapport de l'Ingénieur en Chef de Bresse et Dombes sur le projet du pont en fer et sur celui en pierre qui ont été présénté par le soussigné pour remplacer le Pont des Cordeliers sur la Reyssouze à Pont de Vaux', undated (1789): Dijon MSS pp. 210–13; also further fragments, pp. 216–22. In one of these reference is made to another report on Racle's proposals made by the Dijon Académie 21 February 1789.

54. Perronet, J.R. 'Rapporte de l'Assemblée des Ponts et Chaussées sur un projet de pont en fer proposé par le sieur Racle, architecte, pour être construit sur la rivière de Reyssouze à Pont de Vaux en Bresse, Paris le 10 8bre 1789, signé Perronet': 7 pp. MS in the Ponts et Chaussées dossier (Ref. 38). A covering note states that all of the documents relating to this affair were sent back to Racle on 3 Feb. 1790 except the present one.

55. For William Wilkinson's continental work see, for instance, Dickinson, H.W. *John Wilkinson, ironmaster*, 1914 (particularly pp. 49–53); Chevalier, J. 'La mission de Gabriel Jars dans les mines et les usines Britanniques en 1764', *Trans. Newcomen Soc.*, 1947, *26*, 57–68 (pp. 65–7 on Wilkinson in France); and Henderson, W.O. *Britain and industrial Europe*, 1954 (2nd ed. 1965), pp. 37–46. The company of Le Creusot was formed with a capital of 10 million livres, some British: most of the equipment was of British design and much supplied from Britain.

56. The Société d'Emulation de Bourg was founded in February 1783 and its members included Guyton de Morveau. Racle became an Associate Member 9 January 1785 and submitted various papers including 'Théorie du feu, des fours à bois', 1785 (Dijon MSS pp. 39–42); 'L'Art du tuilier et du briquetier, perfectionné'. 1786 (Dijon MSS pp. 4–56); 'Mémoire sur le commerce locale de la Ville de Lyon . . . et sur la possibilité de la réunion du Rhin au Rhône' etc., undated (Dijon MSS pp. 112–30); 'Un nouveau moyen mécanique pour obtenir une puissant motrice à volonté, à l'exclusion des elements et des animaux', undated (Dijon MSS pp. 518–40). The Society's own records and papers (not seen) are in the Ain archives.

57. Anon, Département de l'Ain, *Journal des Mines*, 1796, *4* (23), 39–50.

58. Dijon MSS Plate IX.

59. Dijon MSS Plates XV, XVI and XVII.

60. Racle, L. 'Mémoire sur un moyen simple et peu dispendieux de garantir les vaisseaux de guerre des intempéries de l'airs pendant la paix, presenté au Comité de la Marine, 1789–90' (two versions plus correspondence): Dijon MSS pp. 60–102.

61. Dijon MSS pp. 274–720 consist of various drafts with titles such as 'Système de la mécanique cycloïdique', 'Mémoire analytique sur les puissances motrices mécanique, et l'exécution d'une roue à mouvements spontannés', and 'Analyse des objections faite contre la possibilité des mouvements perpetués' (all 1789–90). Jarrin states that Henin asked Racle to produce a model of his motor, and that Racle then ordered one to be made by M. Goyffon(!) 'the able director of the horological school'. Obviously this lapsed with Racle's death.

62. *Dictionnaire de biographie française*, 1948.

63. Dufay, C.J. *op. cit.* (note 31).

64. Dormans, Marne, the birthplace of the architect Ledoux, is many miles from Soissons, Aisne. The note referred to offers no clarification.

65. Dartein, F.de. *Etudes sur les ponts en pierre* etc., vol. 4, pp. 103–11 and Plates 5–9.

66. Dufay cites the publications as 'Mémoire sur l'adhérence des maçonneries faites en chaux maigre', Bourg. 1783; ',Mémoire sur l'incohérence des maçonneries faites en gros cailloux et mortier de chaux grasse', Dijon, 1783; and 'Réflexions sur la statistique [*sic*] des voûtes', 1783. The departmental sous-archiviste at Ain (R. Dusonchet) informs me that at least one Aubry paper on rubble masonry exists in the Societé d'Emulation collection.

67. Aubry, N. *Mémoire sur la construction d'un pont de bois de 450 pieds d'ouverture d'un seule jet*, place of publication not known, date variously given as 1789, 1790 and 1791.

68. Dufay gives the title as 'Sur les moyens de garantir les prairies de la Reyssouze, à Bourg, de toute inondation, sans nuire au travail des moulins établis sur cette rivière' and says it was awarded 2nd prize 1 January 1787.

69. P.F.B. Alsop informs me that '*turcies*' was the local name for the *levées* on the banks of the Loire.

70. Aubry, N. *Mémoire sur différentres questions de la science des constructions publiques et économiques*, 4to Lyon, 1790, according to Dufay. Depery referred to a work entitled *Questions académiques*, 4to, Lyon, 1789 which he said contained the prize-winning papers of notes 67 and 68.

71. Gauthey, É.M. *Traité de la construction des ponts*, pp. 115–16. Goiffon's family appear to have come from Cerdon.

72. The first pumping station (Chaillot) began work in August 1781 and the second (Gros Caillou) in July 1786. Both of the Wilkinsons (who provided the 40 miles of iron pipe) were present for the opening celebrations of 1786. The Périers were also involved in Le Creusot.

73. See note 30 for a note on Paine's bridge work prior to his arrival in Paris.

74. *Procès verbaux de l'Académie Royale d'Architecture*, vol. 9, p. 209, (23 July 1787).

75. The committee consisted of Le Roy himself, Jean Charles de Borda, and the Abbé Charles Bossut. Their report is said to have contained a review of all French iron bridge proposals to date (such a review had been one of Paine's main objectives in going to Paris) but no copy can now be traced. Paine eventually saw Perronet in April 1788 but the great man's comments on Paine are unrecorded.

76. Paine arrived in England in September 1787 and began a new round of promotion work but returned to Paris in February 1788. When Vincent's final paper was published in June he rushed back to England to take out his patent (dated 28 August). He had hoped to interest the Wilkinsons in his patent but it was adopted by the Walkers instead. In the next few years Paine's activities undoubtedly helped to give iron bridge discussion the boost it needed in England but contrary to popular legend, there is no evidence that he contributed anything of original engineering value. As one of his close associates (Gouverneur Morris) put it, Paine had 'an excellent pen to write but an indifferent head to think',

77. *Mercure de France*, 26 Jan. 1788 (No. 4), pp. 183–6.

78. Stern, J. *A l'ombre de Sophie Arnould*, 1930, vol.1, p. 234, agreed that Bélanger's design was for a wooden bridge on iron columns but gave a footnote reference to an undated water-colour by Maréchal of a 'projet d'un pont en fer en face le jardin du Roi, par Bellanger [*sic*]'. Hautecoeur, L. *Histoire de l'architecture classique en France*, vol. 4, 1952, p. 65 (and in later articles elsewhere) omitted mention of the wooden design and referred only to the alleged iron one.

79. Other unadopted plans for the bridge are noted by Duplomb, C. *Histoire générale des ponts de Paris*, vol. 1, 1911, pp. 43–4: one for a wooden 'Pont Desiré' by M. Guerne, (a Parisian master-carpenter) in 1788, and another by the entrepreneurial architect P.F. Palloy in 1796.

80. It is interesting that Olinthus Gregory (a pro-Wilkinson writer and an Associate of the Academy of Dijon devoted one of the few biographical entries in his *Cyclopaedia Pantologia* (1813) to Racle, albeit only a few lines.

81. The surviving sketchbook of Reynolds contains drawings from French sources as well as three versions of an iron *platte-band* design for the Longdon aqueduct. Telford, who moved to Shrewsbury in 1786, came under the spell of Wilkinson in 1793 but does not seem

to have been greatly interested in ironwork until February 1795, when the great floods destroyed several masonry bridges in Shropshire but left the Coalbrookdale arch unscathed.

82. Nash advertised for French refugees in the early 1790s and took on Pugin. His first iron bridge had light framed voussoirs with *pendentives*, apparently held together with sleeves. His second bridge had his patented hollow-box voussoirs.

83. Rennie possessed a translation of Vincent's *Prospectus* (auctioned at Sotheby's in 1967) and corresponded with several French engineers. He was a former employee of Boulton & Watt and kept in touch with them continuously: they in turn were in close contact with Wilkinson and the Périers, with the Chaillot works and Le Creusot. Hutton, who assisted Rennie in his bridge calculations, translated Montucla and was no doubt well aware of the French scene: he drew attention to the similarity in plan between the first Telford & Douglass design for a 600ft span iron bridge and Perrault's Sèvres design.

84. James, J.G. 'The cast-iron bridge at Sunderland', typescript as yet unpublished; and 'The cast-iron bridges of Thomas Wilson', paper given at the Newcomen Society, Jan. 1979.

Additional Note

Several *serruriers* have been mentioned. This term is normally translated as 'locksmith' but for the period in question this is misleading. It should be taken to mean a smith working in iron using 'bench' in addition to straightforward 'forge' techniques to produce relatively precise fitted parts rather than ornamental ironwork.

The Early History of Mechanical Harvesting

L.J. JONES

Introduction to the Literature

During the past four or five decades a considerable number of publications dealing with the history of grain-harvesting and other farm machines have appeared on library shelves, but unfortunately this does not signify that the story has been adequately or even accurately told. Some American authors, for example, have followed a chauvinistic compulsion to claim the invention of the reaping machine (in 1831) for their countryman, Cyrus H. McCormick. In so doing they ignored the fact that Patrick Bell's reaper had been working very satisfactorily in Scotland for several years previously. (Incidentally, such authors have also tended to overlook the marked correspondence of the principal features of McCormick's machine with those of Bell's and/or several other earlier British designs). Few would be surprised to learn that a particular offender in these respects was McCormick's own grandson, Cyrus Hall McCormick III,[1] but there were certainly others as well. As late as the end of 1936, for example, an anonymous writer in the American journal *Nature* listed the practical reaping machine as one of 'Twelve Notable American Inventions'.[2] A protest was quickly lodged by Professor James Hendrick[3] of Aberdeen University, but there was no indication subsequently that the anonymous author was prepared to change his stand.

However, while McCormick's machine may not have been the first to operate successfully in the field, there can be no denying that it was an immensely important one in the overall story of the mechanization of harvesting. McCormick thus thoroughly deserves the detailed and scholarly attention he has received in more recent times in such works as W. T. Hutchinson's *Cyrus Hall McCormick* (1930). Nevertheless, it is to British authors that we must turn for satisfactory accounts of a number of earlier (but at most only partially successful) attempts to devise mechanical reapers. Most of these arose either in the south of Scotland or in the adjoining border counties of northern England during the first three decades of the nineteenth century.[4] Some were patented but an equal or greater number were not. Some were actually constructed and tried in the field while others (apparently) advanced no further than the drawing or model stage.

G. E. Fussell, a former civil servant employed at the Ministry of Agriculture and Fisheries in London, has investigated this and related subjects in great depth and has published a large number of excellent books and articles.[5] However, even these are not without their blemishes. In *The Farmer's Tools: 1500–1900* (1952), for instance, Fussell thoroughly confused the Cutting and Reaping Machine patented in England by Ralph Errington Ridley in 1852[6] with John Ridley's South Australian Stripper, invented in Adelaide nine years earlier.[7] The two Ridleys were in fact entirely unconnected, and their machines totally different in both principle and appearance.

Fussell partially retrieved the situation in a half-page entry entitled 'An Early Combine Harvester' in the popular English magazine *Country Life* in 1957[8] — but only partially! Here he identified the Stripper correctly and included an authentic illustration,[9] but he incorrectly claimed that it originated in the 1870s (thirty years too late). He then made two further blunders by stating that (i) the machine *broke* the heads of grain off whole from the stalks, and (ii) the horses which pulled the machine were so positioned that they could not avoid trampling the crop. The truth is that the grain was directly stripped from the ears of the standing crop, and also that the horses walked to the side of the machine to *avoid* damaging the unharvested portion of the crop. That is, the Stripper was designed with side draught in all versions except the very first in 1843 (which, incidentally, was pushed from the rear to achieve the same basic result). In a later book entitled *Farming Technique from Prehistoric to Modern Times* (1966), Fussell included the same illustration of the Stripper, and again neglected to point out that it represented an advanced model rather than the original. However errors such as these are very rare in Fussell's work, and his immense scholarly output stands as a monument to his patience, dedication, and skill in historical research.

Unhappily, Fussell's mistake over the two Ridleys and their separate machines was carried on by other authors long after he himself had corrected it. It was echoed, for example, by the classicist Dr. K. D. White of Reading University in his otherwise fine book *Roman Farming*,[10] published in 1970. Incidentally, a further error in this work is White's claim that a harvester introduced into Britain in 1946 (the Wild Harvest Thresher) operated on the same principle as Ridley's South Australian machine.[11] This appears to be an assumption based solely on the feature of both machines that the straw was left uncut and standing in the field after the grain had been collected. In fact the two were fundamentally dissimilar. The Ridley Stripper threshed the grain directly from the standing crop by means of relvolving beaters, while the Wild machine rubbed and sucked the grains from the ears between rotating baffled discs.[12]

In an earlier book, *Agricultural Implements of the Roman World* (1967), White made a similar mistake[13] in likening Patrick Bell's Scottish Reaper to the *vallus* used on the plains of Roman Gaul in the early Christian era. Once again the two devices are quite unlike in almost every way. On the other hand his treatment of the Roman machine itself is excellent — it will be referred to again in more detail in the following pages.

Erroneous statements concerning Ridley's South Australian machine also occur in other contemporary works published on both sides of the globe. For example, as recently as 1973 the English writer M. Partridge asserted that:

> The first Australian grain-header or 'stripper' was the work of John W. Bull and John Ridley in 1843 . . . Later Bull improved the gathering comb and succeeded in thrashing the grain directly from the standing crop.[14]

In fact Bull did nothing of the kind. Ridley and Bull may well have had roughly the same idea — about the same time but quite independently — for stripping the grains directly from the ears,[15] but it was Ridley alone who introduced a practical machine and demonstrated its capabilities in the field. Bull did nothing more than make known his idea by displaying a model in Adelaide in September 1843.[16]

Perhaps the most distinguished living Australian historian, Professor C. Manning H. Clark, also erred when he claimed (in his *History of Australia*, Vol.3 (1973), p.281) that Ridley's Stripper reaped, threshed *and winnowed* the wheat simultaneously. The truth is that the winnowing function was absent in all Australian harvesting machines prior to Hugh Victor McKay's first design of 1884.[17] To be fair to Professor Clark, however, it should be pointed out that his was only one of a string of errors initiated by the mid nineteenth century works of G. B. Wilkinson,[18] Jacob Wilson,[19] and others.

Thus it would seem that, although a considerable body of writing on the history of grain-harvest mechanization prior to, say, 1850 already exists, there is yet a need to set the record straight in a number of respects. This is broadly the aim of the following pages.

The traditional steps in preparing wheat for human consumption

Cereal crops of various kinds have been deliberately cultivated for food since ancient times. Throughout recorded history wheat (or corn, as it is sometimes called)[20] has been an important part of the diet of Western peoples especially, and today is perhaps more important as a foodstuff than ever before. Traditionally, four separate operations were carried out to convert the growing wheat crop into a form suitable for human consumption:

1. *Reaping.* Generally the stalks of wheat were cut (by men using some form of blade) somewhere between ground-level and about half-way up. (The height of cutting tended to depend upon whether the straw was required for thatching, to be stored as stock fodder, or not to be collected

at all.) Next, the cut stalks were collected up into bundles and bound to form sheaves, several of which were then placed together to make a larger bundle or stook. These stooks were then stood in the open air (usually in the field itself) to allow the stalks fully to dry out, the time required for this being usually some two or three weeks.

2. *Threshing*. In this phase the dried stalks were beaten, trampled, or shaken in some manner to dislodge the grains from the ears. The stalks were then removed to be used for one of the purposes indicated above, or else simply discarded.

3. *Winnowing*. This was really just a cleaning process in which residual foreign matter such as husks, small pieces of straw, or grit were removed. Some kind of sieve was usually employed.

4. *Grinding*. Here the cleaned grain was ground (originally pounded or crushed) to a relatively fine meal suitable for making bread.

Formerly all these operations were carried out by hand. In each case the work was hard, slow, and often wasteful of grain as well. All four tasks were eventually performed by machines of one kind or another, thus relieving men of the drudgery and physical toil involved. Perhaps more importantly, however, the advent of such machines released manpower for other farm work, and so made it possible for much larger quantities of grain to be produced by the same number of agricultural workers.

Grinding was the first of the four operations to be mechanized. The earliest history of flour-milling is obscure, but it is known that, among the Romans at least, the ancient hand-operated querns began to give way to rotary mills turned by donkeys round 200 B.C.[21] Water-power was also harnessed on a significant scale to turn grinding stones from Roman times onwards.[22] The windmill was a medieval invention and, as is well known, increasing numbers of these were erected in north-western Europe and Britain from roughly 1300 onwards and employed for pumping, blowing forges, etc., as well as for driving millstones.[23] By the period with which this paper is principally concerned (the early and mid nineteenth century), European flour-mills had evolved into a fairly standard form. For the actual grinding, flat stone discs with grooved faces were used, driven by a geared shaft from a water-wheel or water-turbine or perhaps by a steam engine, and the use of wind power was noticeably on the decline.

Reasonably satisfactory machines for threshing the grain from the cut stalks became available from the middle of the eighteenth century onwards. However, a design evolved by Andrew Meikle of East Lothian in Scotland in 1786[24] is usually regarded as the first really successful threshing machine, and over the following decades machines of this type came into widespread use on British wheat farms. Effective winnowing machines were also developed in the latter half of the eighteenth century, and were likewise employed in considerable numbers.[25] A detailed discussion of these matters is, however, outside the scope of present considerations.[26]

Mechanical devices for reaping are necessarily mobile, while those for grinding, threshing and winnowing are not. For this and perhaps other reasons the evolution of worthwhile reaping machines proved more difficult and occurred later than that of the others just mentioned.

Traditional tools for hand-reaping grain crops

In ancient times a crude form of sickle is believed to have been the first implement used to cut the stalks of wheat and other cereal grasses. The earliest form probably comprising a crescent-shaped flint fitted into a bone or wooden handle, was in use several thousands of years B.C. and probably originated in the Nile Basin. The iron version was known in England by Roman times, and ultimately took two separate forms which were used in quite different ways. In one form the blade was toothed or serrated up to an inch or two from the point — this was used with a sawing action and continued to be referred to by the original name of sickle. The other, which became known as the reaping hook, had a smooth-edged blade and the stalks were severed by using a slashing action.[27] Much more recently a greatly improved form of cutting tool now known as the scythe was developed, possibly by the Flemish peoples, which allowed the user to adopt an upright stance and to use the power of both hands. The result was much reduced back strain and fatigue, and a rate of cutting almost two-and-a-half times quicker than was possible with the sickle. (See Figures 1 and 2.)

A further improvement was achieved by attaching a structure known as a cradle to the back of the scythe-blade to catch the cut stalks. This further reduced both the time and the physical effort required to gather a crop, especially if it was standing well. A simple form called a bow or bale could readily be made of a piece of bent hazel or a light frame and sacking.[28] Another form, thought to have been introduced in America about 1776,[29] comprised a set of curved bars and was often referred to as a rake. (See Figure 3.) There is however a measure of disagreement over the origin of this style of cradle; the Canadian author Merrill Denison, for example, suggests that it may have been devised earlier by English Quakers and taken to America by them.[30] Evidently further work will be needed to clarify the matter.

Thus it is seen that, although the form of the tools used for reaping changed and indeed improved slowly over the centuries, reaping remained a manual task everywhere until the early nineteenth century — with one notable exception. This occurred in the early agriculture of the Gauls, and it is discussed in detail in the following section.

Figure 1. Reaping with the sickle (from M. Denison, *Harvest Triumphant* (1948), facing p. 32).

Figure 2. Reaping with the common scythe (from M. Denison, *Harvest Triumphant* (1948), facing p. 32).

Figure 3. Reaping with the scythe and cradle (rake form) (from M. Denison, *Harvest Triumphant* (1948), facing p. 32).

An isolated attempt to harvest by machine in the early Christian era

Sufficient references exist to establish that a moderately successful attempt to mechanize cereal harvesting was made in Roman-occupied Gaul between approximately the first and fourth centuries A.D. Nothing is known so far of how or where this method of collecting grain first arose, or why it was subsequently abandoned. It seems very likely, however, that it was in common use in the Gallic provinces for a period of at least three or four hundred years.

The earliest known record of the method occurs in a surviving work of Pliny the Elder, which dates from about the middle of the first century A.D. He describes briefly a special type of cart used in Gaul at the time which collected the ears of grain from the standing crop. Part of Pliny's short note on the subject reads in translation as follows:

> There are various methods of actually getting in the harvest. On the vast estates in the provinces of Gaul very large frames fitted with teeth at the edge and carried on two wheels are driven through the corn by a team of oxen pushing from behind; the ears thus torn off fall into the frame. Elsewhere the stalks are cut through with a stickle.[31]

Another short description of the same device was included in a pamphlet entitled *De Rebus Bellicis*, written by an unidentified Roman author[32] about A.D. 350. However, a more comprehensive account than these two was provided by a later Roman, Palladius, in his *De Re Rustica* which was written in the fifth century A.D. It states that:

> In the plains of Gaul they use this quick way of reaping, and, without reapers, cut large fields with an ox in one day. For this purpose a machine is made, carried upon two wheels; the square surface has boards erected at the sides, which, sloping outwards, make a wider space above; the board on the fore part is lower than the others; upon it there are a great many small teeth. wide set in a row, answering to the height of the ears of the corn, and turned upwards at the ends; on the back part of this machine, two short shafts are fixed, like the poles of a litter; to these an ox is yoked, with his head to the machine, and the yoke and traces likewise turned the contrary way: he is well trained, and does not go faster than he is driven. When this machine is pushed through the standing corn, all the ears are comprehended by the teeth, and heaped up in the hollow part of it, being cut off from the straw, which is left behind, the driver setting it higher or lower, as he finds it necessary; and thus, by a few goings and returnings, the whole field is reaped. This machine does very well in plain and smooth fields, and in places where there is no necessity for feeding with straw.[33]

After Palladius, the machine (called a vallus by the Romans) is not mentioned again in the surviving literature until the eighteenth century, when interest in the prospect of harvesting grain mechanically began to revive. This strongly suggests that the Gallic method of reaping may have died out soon after Palladius' time.

Unfortunately, none of these three early accounts were illustrated, or if they ever were the illustrations have long since been lost. As a consequence, the precise structure of the vallus remained a matter for speculation over the following centuries. That notwithstanding, one might expect that some attempt would have been made within a reasonable period to follow up the promising beginning made by the Romans, but such was not the case. The tantalizing descriptions of the Gallic machine provided by Pliny, Palladius, and the anonymous author were either forgotten or ignored for thirteen centuries and the impetus was lost. Reaping reverted to the use of the hand-held reaping-hook and similar implements, and even these changed very little with the passing centuries.

During the eighteenth century translations of various old documents, including those mentioned above, were made and began to circulate. As a result:

> the ghost of the Gallic reaping cart emerged repeatedly in the agricultural literature before 1800, . . . [but] without any effects.[34]

Also during the eighteenth century several attempts were made to interpret the old Roman description diagrammatically. The drawing made by the French engraver Lasteyrie, which is shown in Figure 4, seems to be fairly typical of that time.

Speculation concerning the basic form of the vallus was finally brought to an end as recently as May 1958, when a chance archaeological find in Luxembourg provided the first authentic picture of the machine with recognizable detail. At Montauban-Buzenol, near Arlon, a third-century stone tablet was unearthed which depicted (in bas-relief) what is unquestionably one of the Gallic reapers in use at or near that time (see Figure 5).

From a comparison of this tablet with Lasteyrie's conceptual drawing it is seen that the actual machine was more developed than had been expected. The spoked, chariot-like wheels had not been anticipated, and (at least according to G. E. Fussell)[36] the appearance of the frontal spikes suggests that they were made of iron rather than wood. More recently still, however, K. D. White has proposed that two different styles of machine were in fact used in Roman-occupied Gaul.[37] On etymological grounds White concludes that the vallus referred to by Pliny had a shallow body resembling a basket lying on its side, whereas the machine described by Palladius was of a type similar to that shown in Lasteyrie's drawing in Figure 4. Because of its different body-shape, White prefers to call the latter a *carpentum* rather than a vallus, but nowhere does he suggest that the two were anything more than variants on the same basic theme.[38]

After the discovery of the stone tablet at Montauban, however, at least one question still remained to be settled — that of the performance of the machine as compared to reaping by hand. Within a few years of the find this was carefully tested by a group within the German Society of Agricultural Engineers, who:

> reconstructed the stripper-reaper[39] from the elements they could identify . . . They took care to use construction methods that would have been available at the time the original was developed. Field tests revealed this device to be about three times faster than cutting wheat with a sickle, then flailing and winnowing it — a technique used even today in some less developed nations.[40]

Figure 6 shows this reconstruction.

As a matter of fact, this report is somewhat misleading since the vallus did no more than collect the heads of grain. A winnowing process and probably some threshing of the heads as well would have been necessary after the machine had done its work. It would have been fairer to compare the time for machine-reaping to that required for hand-*cutting* only without including 'flailing and winnowing' as was evidently done by the German Society. But even so, there can be little doubt that the device did indeed enable the crop to be gathered at a faster rate than by men using

Figure 4. Conjectural drawing of the Roman vallus from C. P. de Lasteyrie, *Collection de Machines* (1823) (from a copy included in J. C. Loudon, *An Encyclopedia of Agriculture* (1825), p. 26).

Figure 5. A section of the Trevieren[35] Tablet, found at Montauban in 1958 (from *Agricultural Engineering* (October 1973), p. 51).

Figure 6. A modern reconstruction of the vallus by the German Society of Agricultural Engineers (Max-Eyth-Gessellschaft zur Förderung der Landtechnik) (from *Agricultural Engineering* (October 1973), p. 51).

Figure 7. William Pitt's reaping machine design of 1786 (from Arthur Young's *Annals of Agriculture* (1787), p. 161).

sickles or hooks. The fact that it remained in use for three centuries or more in itself provides considerable support for this opinion.

However, there is no specific information in any of the early Roman accounts to explain how the vallus arose in Gaul in the first place, or why it was later abandoned. According to Palladius the machine was used to save manpower,[41] from which it may reasonably be inferred that it was introduced to cope with a general labour shortage on the farms. K. D. White has gone further by suggesting that:

> a heavier demand for wheat for the [Roman] Rhine armies,[42] and shortage of seasonal labour would be the most likely conditions for such a development.[43]

Since slave labour was not available in the Gallic provinces to any marked degree, White's explanation seems plausible. However, finding a convincing explanation for the machine's disappearance — apparantly in the fifth century A.D. or soon afterwards — is more difficult. H. A. Beecham and J. Higgs suggest that it was because:

> The method was wasteful, especially of straw; but at the time it [the vallus] was used there was much land and the crops were thinly sown; also the population was small and labour scarce. Later, the population increased, labour became more plentiful, and greater

yields were required. It was probably for these reasons that the machine was abandoned.[44]

An explanation along broadly similar lines was that offered by the American author Lynn White, Jnr., who said:

> the Gallo-Roman grain harvester was ingenious, yet was unknown during the Middle Ages. Its defect in medieval eyes was that it wasted the straw, and was therefore appropriate only to a regime of cereals little related to animal husbandry . . . one of the great agrarian advances in Northern Europe during the early Middle Ages was the development of a much more intensive and productive agricultural system combining grain with stock raising. In such a complex, straw had great value, and the Gallo-Roman harvester was worthless.[45]

Unfortunately, neither of these explanations is acceptable. Beecham and Higgs are simply wrong in supposing that any marked increase in population occurred at about the time the vallus disappeared. Indeed, there was clearly a substantial decrease in the region's population due to the withdrawal of the Roman army of occupation. On the other hand, Lynn White's theory accounts for the demise of the vallus at the wrong time. The last surviving account of the machine (that of Palladius) was written several centuries before the early Middle Ages when the new style of agriculture is supposed to have arisen. In addition, Lynn White's (and also Beecham and Higgs') argument that the machine became worthless merely because the wheat straw was required to supplement the natural diet of stock, is weak and inconclusive.

The point Lynn White seems to have missed is that, shortly after Palladius recorded his account of the vallus in the fifth century, the Romans lost effective control of Gaul as a result of the Frankish invasions. The former Roman occupiers of land tended to stay on, but their influence rapidly declined as the Franks established themselves. With the withdrawal of the Roman armies the demand for grain also undoubtedly declined, thus leaving the way open for the native Gallic farmers to revert to their old ways of small-scale cropping (on smaller fields), less specialization, and more insular living. In these circumstances it is conceivable at least that the vallus simply became unnecessary to them, and was quietly put aside.

The first modern attempts to devise mechanical reapers

A renewed interest in the vallus seems to have occurred first in England during the second half of the eighteenth century. About this time, popular interest in the contents of surviving ancient texts was stimulated by the publication for the first time of English translations of some of them in magazines and readily available journals. A number of these dealt at least in part with scientific or technical subjects — a circumstance which probably increased rather than diminished reader interest, since such topics were the fashionable ones of the day. Previously, of course, the contents of such ancient texts had been available only to the favoured few whose education enabled them to read the original Latin or perhaps Greek, and unfortunately these were generally not the kind of people to whom technical matters were either interesting or comprehensible.

A couple of minor attempts during the 1780s to follow up the ancient Gallic method of gathering grain crops were apparently the first made since Roman times, as is indicated by the following:

> The earliest proposal for a mechanical reaping machine in Britain, appears to be that described in the *Annals of Agriculture* by Arthur Young, and published in the year 1785, wherein it is stated that the Society of Arts, in the year 1783, offered a premium of a gold medal or £30, for a machine for mowing or reaping grain crops. This was taken up by a Mr. Capel Lofft in 1785, and by Mr. William Pitt in 1786, both of whom seem to have acted on the information of Pliny, and, in fact, combined the principle of 'rippling' with that of *reaping*, leaving the straw to be collected at leisure; — thus showing that improvement in reaping machines during the 1800 years which had elapsed, was scarcely appreciable.[46] [See Figure 7.]

It is perhaps significant that the offer by the Society of Arts in 1783 followed the Society's publication earlier in the same year of a translation of Pliny's first-century description of the vallus.[47] Soon after these (apparently unsuccessful) efforts by Lofft and Pitt, however, attention was diverted to a different approach in which the emphasis was laid on developing cutting devices of various kinds. In this the aim became one of more or less duplicating mechanically the action of the sickle or scythe in the hands of a man.

The first English patent was granted to Joseph Boyce in 1799[48] for a type of mowing machine. Boyce's specification drawing (see Figure 8) shows that he proposed to cut the crop with a set of scythe blades attached to a circular disc rotating in a horizontal plane close to the ground.

Robert Meares secured a patent the following year for a 'Machine for Cutting, after a new Method, Standing Corn, Grass, And the like, and for Making Reed',[49] but there appears to be no record showing that Meares' design was ever actually made or tested. In any event, it was really nothing more than a single set of long-handled clippers mounted on a pair of wheels, which was to be pushed forward by one man while the clipper was actuated manually by another — presumably walking alongside.

Around 1802, two machines of contrasting types were reportedly tried by an unnamed French inventor, but evidently unsuccessfully in both cases.[50] In 1805 a patent was granted to a Mr. T. J. Plucknett of Deptford[51] for a design based on a notched circular cutter (see Figure 9), and an improved version of this was introduced in model form by

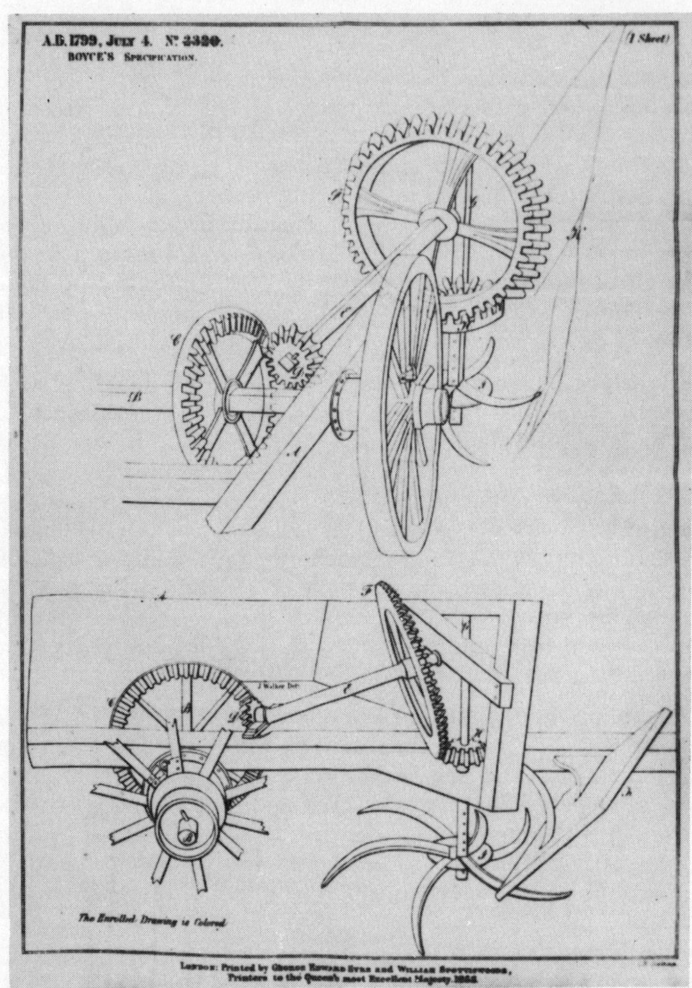

Figure 8. Patent drawing for Joseph Boyce's reaper, filed in 1799 (copy supplied by H.M. Patent Office, London).

Figure 9. Plucknett's reaper design of 1805 (from G. E. Fussell, *The Farmer's Tools: 1500–1900* (1952), Fig. 48 and reproduced here by permission of the Hutchinson Publishing Group).[52]

Figure 10. Mr. Gladstone's reaping machine design, first proposed in 1806 (from Robert Brown of Markle's *Treatise on Rural Affairs* (1811), Figs. 39–43).[55]

Mr. Gladstone of Castle Douglas in Kirkcudbrightshire, Scotland in 1806 (see Figure 10). Gladstone's improved design incorporated a mechanism for gathering up the cut stalks, but when a full-sized machine was eventually made and tried in 1811 it failed to work properly. G. E. Fussell mentions another interesting attempt at reaper design in 1811 by a

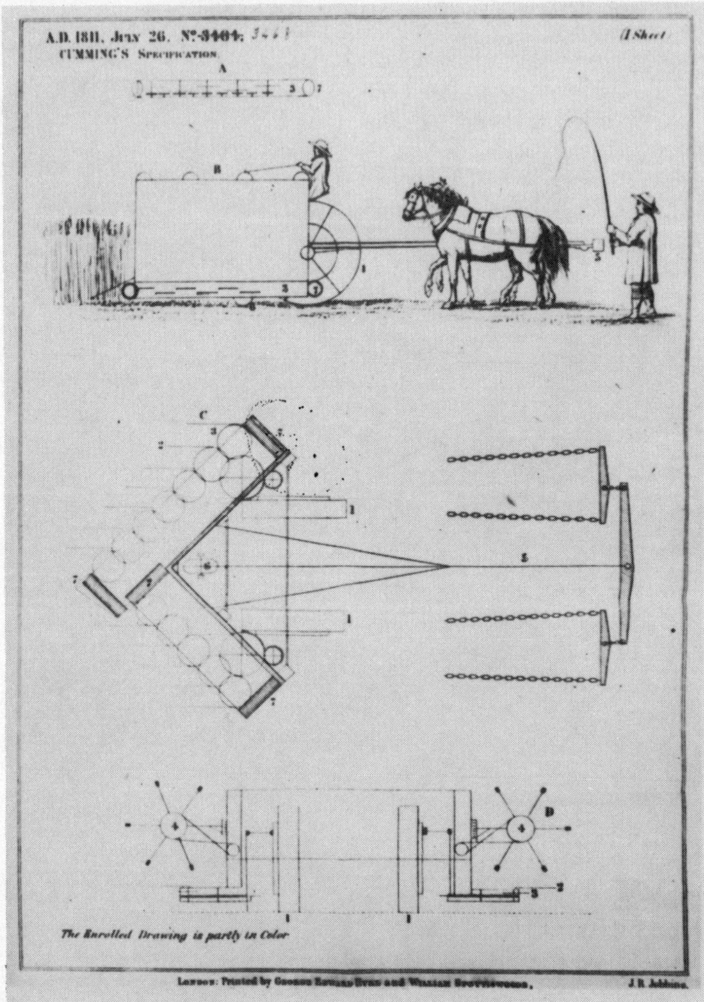

Figure 11. Patent drawing for Donald Cumming's reaper of 1811 (copy supplied by H.M. Patent Office, London).

Northumbrian farmer named Donald Cumming,[53] while H. A. Beecham and J. Higgs refer in their booklet to a Northumbrian design supposedly tested in 1812 by one John Common.[54] From the similarity of the names and noting that both machines were claimed to have arisen in Northumberland at near enough the same time, it seems not improbable that there was in fact only one machine, and that 'Cumming' and 'Common' were one and the same man. If so, the correct name is Donald Cumming, since a patent for a reaping machine was registered in that name in 1811.[56]

Figure 12. Robert Salmon's reaper, *c.*1808 (from G. E. Fussell, *The Farmer's Tools: 1500–1900* (1952), Fig. 48 and reproduced here by permission of the Hutchinson Publishing Group).

Further, it must be said that Cumming's unusual triangular design (see Figure 11) is now of considerable interest, since it appears to have been the first to include a reel to sweep the crop towards the cutters, and the first also to feature endless moving cloth belts to transfer the cut grain to the side of the machine in a tidy swathe. As will be shown later in this paper, both were subsequently important features of the successful designs of Patrick Bell, C. H. McCormick, and perhaps others.

However, the most bizzare attempt at reaper design of this early part of the nineteenth century must surely be that of James Dobbs in 1815.[57] Mr. Dobbs was an actor employed at a Birmingham theatre and, having actually constructed a machine to his own plan, he then tried to publicize it by demonstrating its capabilities as part of his theatrical act. According to Fussell,[58] Dobbs even went to the extraordinary lengths of setting up a small patch of growing wheat on the stage of the theatre for the purpose.

Dobb's patent specification reveals, however, that his machine was in fact little more than a frame supported on two wheels at one end and attached by straps to a man's shoulders at the other, on which was mounted a hand-operated mechanism to turn a series of cutting discs mounted between the wheels. From his drawing it appears that there were six such cutting discs, and hence it is very doubtful indeed that the device was a truly workable one — obviously the power required to turn the mechanism would have been well beyond the ability of a single man to provide. But at the same time Dobbs' odd conception embodied one feature of interest, viz. a set of tapered crop-dividers mounted ahead of the cutters. This was a feature not found on any previous design, and (like the reel and the cloth belt noted earlier) one which was to reappear on many others later.

The reaping machine of Mr. Robert Salmon of Woburn (Figure 12) is believed to have been introduced around 1808, but it was not patented and its performance appears to have been generally unsatisfactory. However, this design again included a gathering mechanism of sorts, and it was also apparently the first to cut on the clipping rather than the rotary principle.

In the meantime the first reaping machine patent in America had been granted to T. I. Hawkins of New Jersey in May 1803, and several other American designs were registered in the following years. None were notably successful. Indeed, none of the reaper designs discussed so far — British or American — could be described as even tolerably effective in the field, and certainly none of them ever came into general use.

Nowadays there is a considerable measure of agreement among writers on this subject that the most promising of the pre-1820 British cutting machines was one constructed by James Smith of Deanston in Perthshire, Scotland, and first introduced in 1811 (Figure 13). However, Smith was not satisfied with this first model, and he proceeded to modify the design many times for several years afterwards. Eventually it won for him two separate prizes when it was exhibited at agricultural shows in Scotland. In each case Mr. Smith received a piece of plate valued at fifty guineas — a substantial sum for that time — but the machine failed to win general approval and adoption by the farmers. According to the mid nineteenth century author James Slight,[59] the reasons for its lack of appeal were that it was too heavy, too awkward to manoeuvre, and too expensive.

Other attempts followed in quick succession, of which those by Kerr of Edinburgh, Scott of Ormiston (in 1815), and especially Joseph Mann of Raby (Cumberland) deserve some mention. Mann's design was embodied in a model in 1820, and this was followed by a full-sized machine in 1822. Unfortunately it failed to perform adequately when tested in the field. Some years later Mann resumed his efforts, and in 1832 he produced a machine which (according to James Slight) worked fairly satisfactorily (Figure 14).[60] Jacob Wilson went further, claiming that the machine 'cut a breadth of $3\frac{1}{2}$ feet and with *one horse* could with ease accomplish 10 acres

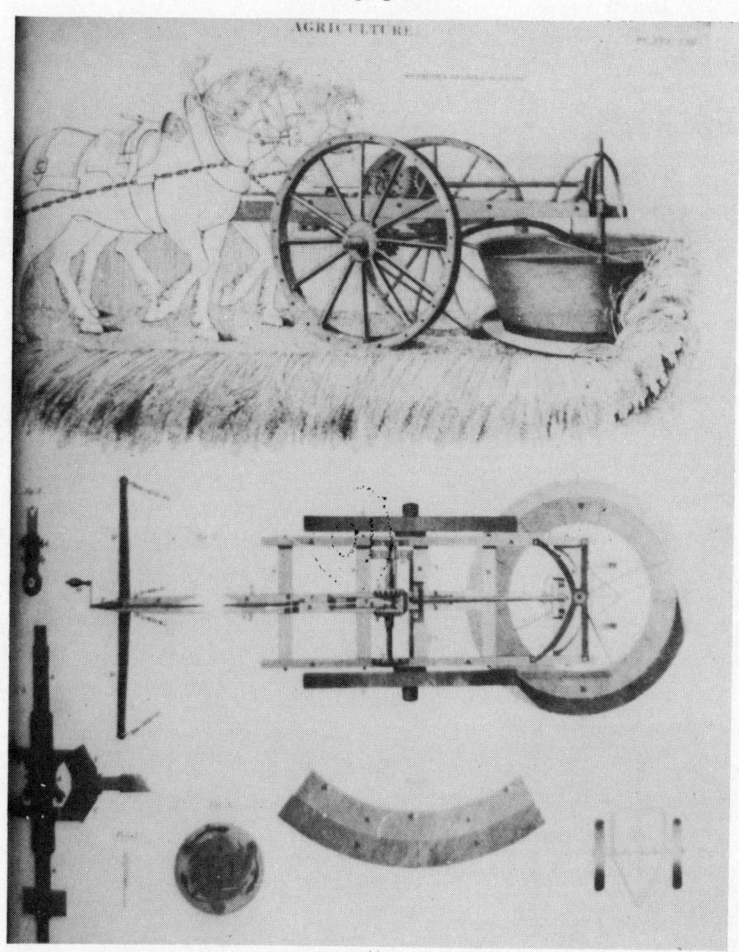

Figure 13. Smith of Deanston's reaping machine (from *Encyclopedia Britannica*, Vol. 2 (1842 edition), Plate 8).

per day'.[61] It was, however, a mechanically complicated device, and as a result subject to frequent and troublesome breakdowns.

In 1822, Henry Ogle of Rennington, Northumberland, devised a much simpler style of reaper, employing a scissor-type cutter at the front and a reel to sweep the crop onto the cutter-bar (Figure 15). Of course, both these elements had appeared previously — the oscillating cutter in Salmon's machine and the reel in Donald Cumming's unusual triangle-shaped device of 1811 — but Ogle's was the first design in which they appeared in combination. (Incidentally Cumming was also Northumbrian and farmed only fifteen miles or so from Rennington where Ogle earned his living as a schoolmaster.)

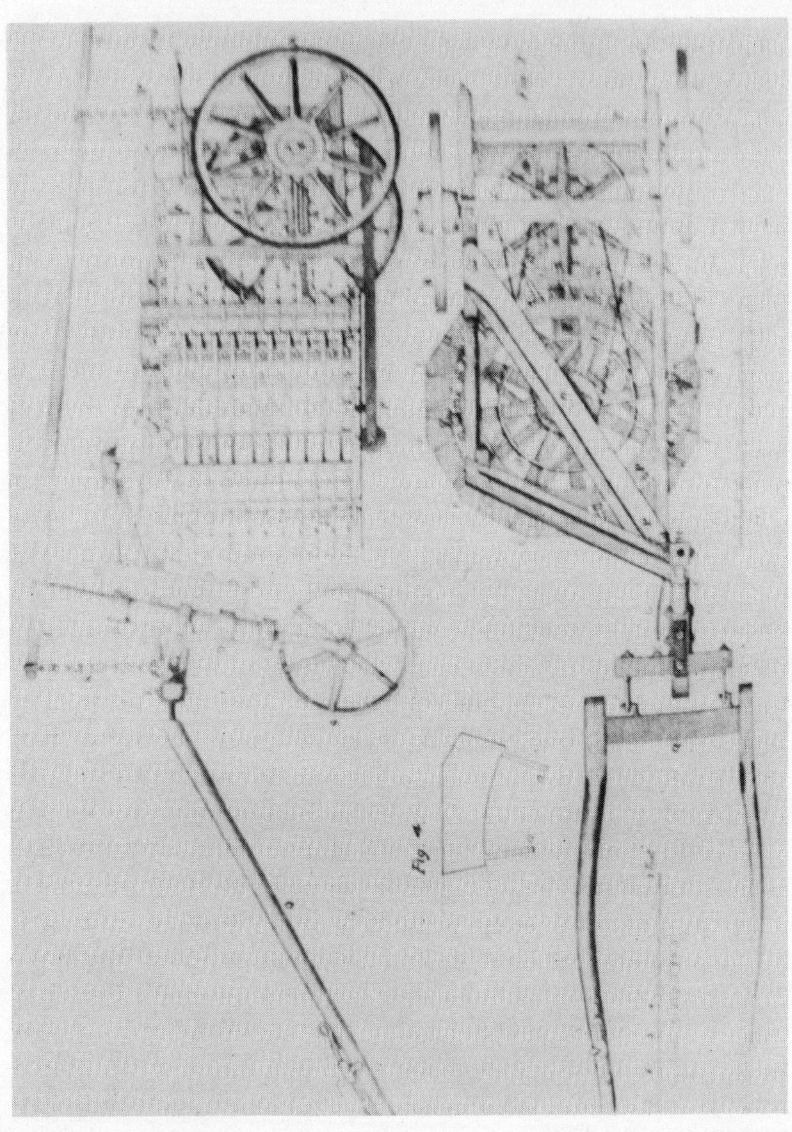

Figure 14. Joseph Mann's reaping machine (from *Quarterly Journal of Agriculture* (1832–4), p. 255).

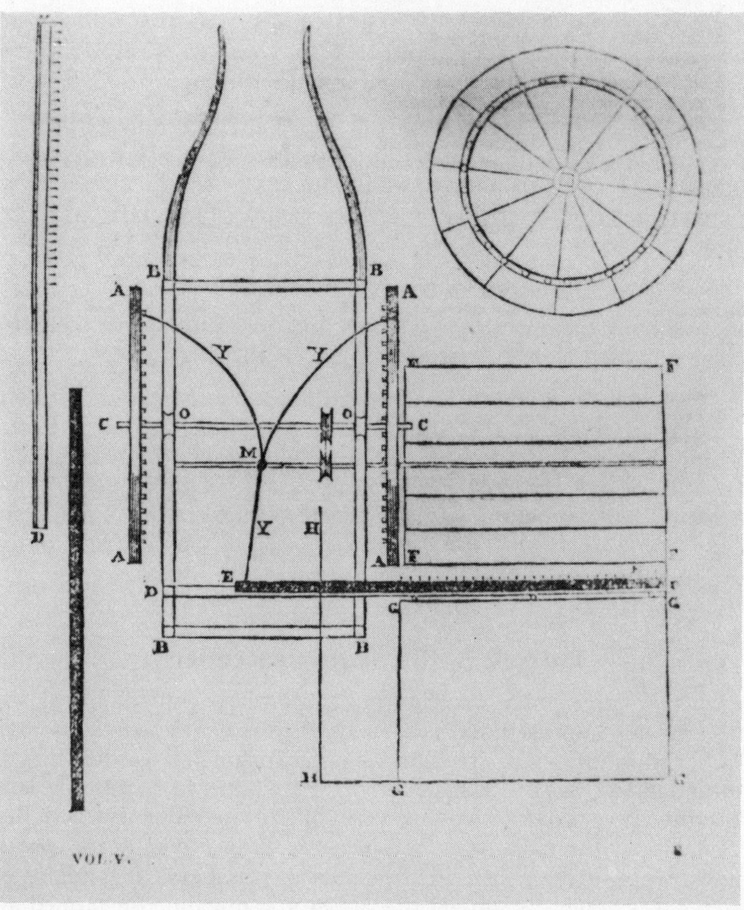

Figure 15. Henry Ogle's reaper design of 1822[62] (from *Mechanic's Magazine* (12 November 1825), p. 49).

Merrill Denison suggests that Ogle's was the first really practical reaper, and claims that it was capable of harvesting up to fourteen acres of wheat in a day.[63] Denison's opinion appears to be based upon those of the nineteenth-century writers, James Slight and Jacob Wilson, mentioned previously, and if she is right, the machine must indeed have worked quite effectively. Even a good man with a scythe, it must be remembered, could manage no more than two or two-and-a-half acres in a day.[64]

Jacob Wilson also recorded that Ogle's design was further improved when:

in the same year [1822], an apparatus for collecting the corn into sheaves was added by Messrs. Brown, of Alnwick.[65]

Apparently, this feature did not reappear on any other reaping machine until the early 1860s. James Slight drew attention to a different but equally interesting point when he remarked upon:

> a very curious coincidence in the almost perfect sameness, in every point, between Ogle's machine and one of the [later] American reapers, Mr. M'Cormack's [McCormick's], . . . the similarity is so perfect that the description of either would suit equally well for the other.[66]

But despite its promise, Ogle's reaper:

> met with coolness from the farmers and opposition from the working classes, which caused it eventually to be thrown aside.[67]

The abundance of cheap agricultural labour in England at this time (due in part at least to the return of large numbers of men from the Napoleonic wars) had resulted in a general lack of interest in the manufacture or use of any such machine, no matter how encouraging its performance. After the publication of the description of the machine in the *Mechanic's Magazine* in late 1825 (which was written by Ogle himself), nothing more was heard of it.

Patrick Bell's Scottish Reaper

Many of the claims made on behalf of some reaping machines designed prior to 1825 are confused and confusing. As already shown, some writers have believed that the reapers of Smith of Deanston and/or of Henry Ogle were practical, or worked successfully. Such claims are arguable, and in any case the criteria themselves are very difficult to define. But whether or not the claims are true, the fact remains that none of the pre-1825 machines were ever made in substantial numbers or widely used. Nowadays it is generally agreed that a reaping machine developed about 1827–8 by Patrick Bell marked the beginning of the modern era of mechanical harvesting. Bell's original machine has been preserved for posterity on that basis, and it is now displayed at the Science Museum, London. Why this should be so is not at all clear, since Bell's reaper was likewise never made or used in significant numbers, and its effect upon the traditional practices of reaping grain crops by hand was negligible both in Britain and elsewhere. It is also curious that Bell's invention is officially dated 1826 (by the Science Museum),[68] since Bell himself made it clear that he:

> fixed upon the plan now successfully in operation . . . in the summer of 1827 . . . and first experimented with a full-sized machine at the harvest of 1828.[69]

At the time he designed his machine Bell was a young theology student; afterwards he became the resident clergyman at Carmyllie in

Forfarshire, Scotland. As far as is known he received no formal training in mechanics, and his interest in devising a machine for reaping arose solely from his experience of helping with the harvest on his farther's farm.

As indicated in Figure 16, the machine was pushed into the crop by horses yoked to a long pole at the rear, and like its predecessors it was a cutting or mowing device which cut the stalks of wheat close to the ground. The cutting apparatus consisted of a fixed set of triangular blades (bolted rigidly to the frame), and a moving (pivoted) set of similar shape actuated by a transverse rod. A reciprocating motion for this rod was provided by a crank mechanism driven from the main ground-wheels, thus producing a clipping action in the cutters like that of a series of pairs of scissors. The entire cutting system was supported on a separate pair of small wheels mounted on the frame ahead of the main wheels, so that the cutters could at all times be maintained at the desired cutting height.

Above the cutters was a set of revolving wooden vanes (i.e. a reel) so placed as to sweep the crop firstly into the cutters, and then backwards over the cutting mechanism onto a sloping and transversely-moving canvas apron. This was simply an endless belt of canvas passing around a pair of sloping rollers which were driven from the ground-wheels by a combination of gears and chains. (Cf. Donald Cumming's use of similar cloth belts in 1811 in Figure 11). Provision was also made for reversing the drive to the rollers, thus enabling the cut stalks to be laid in a neat swathe on whichever side of the machine might be desired, ready for collection and binding. (See Figure 17.)

All the major features of Bell's machine had appeared previously in other designs. The oscillating bar and scissors-like cutting action had been foreshadowed by Salmon of Woburn (1808?) and Henry Ogle (1822), whilst the reel and the travelling apron had already been seen in Cumming's adventurous design of 1811. Bell publicly insisted, however, that he had not seen or even read about any of the earlier reaper proposals (except that of Smith of Deanston) before constructing his own. It was therefore not possible, he said, that he had been influenced in any way by any of them.[71]

Compared to the other promising reapers of the time (e.g. those of Smith and Ogle), perhaps the most advantageous features of Bell's design was the transversely-moving apron. By laying out the stalks in an orderly fashion alongside the line of the machine, it greatly reduced the work of gathering and binding into sheaves after the crop was cut. The reversing arrangement for the apron drive was also a useful feature — and incidentally a novel one as well at this particular time. Apart from this, the merit of Bell's design resided in the skilful combination of its various elements.

As a matter of Christian principle, Bell declined to apply for a patent[72] and announced that he wished the machine to remain freely available to all who might care to make or use it. Years later, he reported the short-term result of this action as follows:

Figure 16. The original Scottish Reaper, constructed by Patrick Bell[70] (from a photograph of it as now displayed at the Science Museum, London).

Figure 17. Diagrammatic representation of Bell's reaper, showing its mechanism (from J. C. Loudon, *An Encyclopedia of Agriculture* (1831), p. 425).

In consequence of the success of the machine in 1828 and 1829, [when it was demonstrated to selected audiences] there were four or five machines made for private parties, but owing to the ignorance and incompetency of the workmen they were of very bad workmanship. They were perpetually breaking, and after two or three seasons they fell into deserved disuse . . . out of some twelve or eighteen machines made . . . about this time, I am not sure that any one of them continued to work above a few years.[73]

In the same account, however, Bell went on to show that, when properly constructed and used, his machine gave no such trouble:

I can state . . . that the most complete success attended the one of which I took the chief charge. For thirteen successive seasons the whole crop on my father's farm [at Auchterhouse, Scotland] was reaped by it in a very satisfactory manner, no season having occurred during that time in which the corn was so much lodged or twisted as to prevent it being successfully operated upon.[74]

The machine was afterwards taken to the farm of Bell's brother at Inchmichael[75] where it worked satisfactorily for something like a further twenty-five harvest seasons. In 1868 it was transferred to the Science Museum in South Kensington, London, for preservation as an historic technical artefact, and it has remained there on public display ever since. An improved cutting apparatus had been substituted for the original by this stage (see Figure 16), but in all other respects the machine was substantially the same as when Patrick Bell first tested it forty years before.

Bell judged correctly when he suggested that the public reputation of his reaper had been harmed by the unsatisfactory performances of copies poorly made by others. But on the other hand, some of these machines were well-constructed and gave good service; Fussell has recorded, for example, that in the year 1832 ten machines based on Bell's design successfully cut 320 Scotch acres (or about 260 Imperial acres) on British farms.[76] Clearly Bell's reaper was a success as measured by its ability to cut and deliver the corn ready for binding — *provided* it was correctly constructed and managed. It is certain that a few machines based on Bell's pattern found their way overseas during the early 1830s, but whether they performed adequately in their new locations is not recorded. Four are known to have been taken to the eastern U.S.A., and the possible significance of this for the development of American reapers at about the same time and in the same region should not be overlooked (see further discussion later). One or two were sent to Poland, and one at least was also shipped to Van Dieman's Land.[77] What can be asserted with confidence is that the impact of these machines upon the traditional harvesting practices of all those regions was entirely negligible.

Bell and his mechanical reaper were likewise ignored in Britain until after the Great Exhibition held in London in 1851. Bell himself was not an exhibitor, but a display of American reapers designed by McCormick and Hussey (see later sections) attracted considerable attention from both farmers and potential manufacturers.[78] Bell responded by displaying his machine at the annual show of the Highland and Agricultural Society at Perth in Scotland the following year (1852), where, according to him, 'it passed off then for an improved machine'.[79] In fact, no changes of any importance (except for the improved cutter already mentioned) had been made since its introduction twenty-four years previously. Shortly afterwards, the old Scottish Reaper was taken up by the English implement-manufacturing firm of Crosskill, but by then the relatively sophisticated American models were beginning to be imported into Britain in considerable numbers. Despite an almost total lack of development over the previous two-and-a-half decades the Crosskill-Bell reaper held its own for a few years against the American imports in reaping competitions held up and down the country.[80] Eventually the technical lead of the Americans proved to be too great, and the manufacture of machines on the Bell pattern was discontinued.

Early reaping machine developments in America

The agricultural and social conditions of England and America differed considerably in the early nineteenth century, as did also their styles of wheat farming. Agricultural labour in Britain was in over-supply and therefore cheap,[81] but not so in America. Traditional methods of harvesting were deeply embedded in the cultural life of rural Britain, whereas Americans necessarily contemplated change more readily. British wheat farms were (and indeed still are) usually small in area while those on the American Midwest plains were (and also still are) generally very large indeed. These and other indigenous factors had a significant influence upon the kind of progress towards mechanization in agriculture made in the two regions.

In the U.S., serious attention was given to the possibility of harvesting grain mechanically as early as 1803, when T. I. Hawkins obtained the first American reaper patent (see above). However, this and a number of other American reaper proposals put forward during the following twenty-five years were as unsuccessful as the British ones of the same period. The first American design of real promise was patented by William Manning in 1831.[82] This was a pushed machine employing a horizontally reciprocating cutter-bar set at a level to cut off the heads of grain only.[83] It was also apparently the first American machine to make use of a crop-divider to assist the cutting action.[84] The intended advantage of this style of harvesting (known as heading) was that, if convenient, the severed heads of grain could be threshed immediately after collection, but unfortunately:

> the damp weather conditions [of the central-west] . . . made it difficult to thresh grain in the field because it had not had ample time to dry out or cure. As a result the threshed grain spoiled in the bin and became worthless for grinding into flour.[85]

Manning's design was not the answer for the mid-west wheat growers; their need was for a machine which would cut the stalks low down, so that the grain could be dried in stooks before threshing. Two machines designed to cut in this way were introduced separately in 1833 and 1834, by Obed Hussey and Cyrus H. McCormick respectively. Hussey patented and marketed his design first, but McCormick was the earlier of the two to demonstrate publicly the effectiveness of his machine in the field.

Hussey's Reaper

Little is known of Obed Hussey's life until he reached the age of about forty,[86] except that he spent a number of years at sea. How he became interested in the problem of harvesting grain mechanically is unknown. However, it has been established that by 1833 he was living on the farm of Judge A. Foster near Cincinnati, Ohio, and it was there that he built his first reaper. Encouraged by the results of preliminary trials he had conducted privately on the farm, Hussey exhibited his machine at the Hamilton County Agricultural Society Fair in July 1833. Several components broke during demonstrations given at the Fair, but the machine performed well enough overall to gain a written commendation from the Society.[87] Hussey took out a patent for it in the following December.

Despite the fact that it was little more than a simple mowing machine, Hussey's reaper possessed two very meritorious features, viz. an excellent cutter, and a simple mechanical system. Thirty years after its introduction Jacob Wilson stated that it was:

the original from which three-fourths of the machines of the present-day [1864] have been copied.[88]

However this comment of Wilson's referred to the situation in Britain at the time, and in any case was almost certainly an exaggeration. Nevertheless it does indicate that Huusey's machine was of considerable significance in the overall development of mechanical harvesting.

The cutter on Hussey's original machine was an improvement on the type used earlier by Patrick Bell. The blades were of a similar form (acute-angled and shaped like an arrowhead), about seven or eight inches long, and spaced about three inches apart. But unlike Bell's cutter (which operated on a pivoting principle) Hussey's moving blades were given only a pure lateral motion by the rigid bar to which they were attached (see Figure 18). In addition, in most if not all of Hussey's machines the edges of these moving blades were serrated (like a saw) and also constrained to move in slots in projecting bosses of a similar shape to the blades themselves. These bosses simultaneously acted as a set of crop-dividers, as sheaths for the moving blades when the latter were at rest, and as an aid to cutting when the blades were in motion by supporting the stalks at the moment of impact of the cutting edges. The bosses thus provided a momentary extra stiffness in the stalks and so ensured a cleaner cut; at the same time, of course, the bosses served as the fixed set of blades themselves.

The machine was drawn from the front by two horses walking to the side of the cutters, i.e. side-draught, but this was no new feature at this stage. (As previously noted, side-draught had appeared in several earlier British designs, e.g. those of Gladstone (1806), Ogle (1822), and Mann (1820).) The cut corn fell backwards onto a platform behind the cutters, where it was raked into bundles by a man standing on the moving platform and then dropped behind the machine. A second man guided the machine through the crop by riding on the back of the near-side horse.

In light crops Hussey's cutter worked particularly well, but the machine as a whole was difficult to manoeuvre in small fields due to its weight and the size of its large ground-wheels.[89] Further disadvantages were:

1. Friction forces in the cutters were high because of the dry sliding action of the moving blades in their slots. This necessitated fairly frequent stops to change to fresh horses and to re-sharpen the blades.

2. In heavy crops the cutters had a tendency to choke unless the machine was operated at an uncomfortably fast pace. According to Jacob Wilson[90] the horses sometimes had to be driven at a brisk trot to keep the cutters clear. This, of course, was tiring for both animals and men, and again led to regular stoppages.

3. There was no mechanism to deliver the cut stalks in a swathe, and so an extra man was required to remove them from the platform. In addition, the man concerned was exposed to considerable danger, balancing on the moving platform in close proximity to the cutters.

4. No reel was provided to bend the stalks towards the cutters. For some crop conditions the man on the platform had to use his rake to do this, as well as trying to perform his normal task of raking away after cutting.

To a degree these disadvantages offset the machine's admirable simplicity of design, and were the source of some criticism. Hussey nevertheless began making reapers for sale in 1834, and ensured that his machine became the first to enter the American market. Sales were slow, however, throughout the remainder of the 1830s because of doubts on the part of most farmers concerning the benefits of mechanical reaping generally. Hussey continually modified and improved his design, but he failed to gain the widespread approval for his machine for which he had hoped. During the 1840s he did a little better, but still lost most of the small market of that time to his rival McCormick whose price was the more attractive.[91]

By the time Hussey exhibited his reaper at the 1851 Great Exhibition in London, its appearance had changed very much from what it had been initially. The two large ground-wheels and other smaller wheels (see

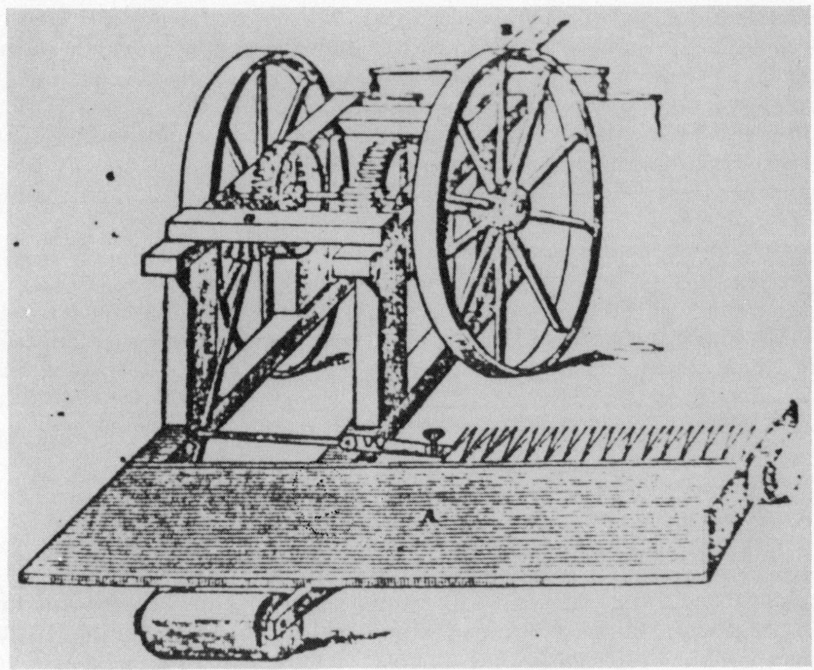

Figure 18. Obed Hussey's original reaper (from *Quarterly Journal of Agriculture* (January 1854), p. 197).

Figure 19. An improved Hussey Reaper of about 1851 (from *Journal of the Royal Agricultural Society of England*, Vol. 12 (1851), p. 613).

Figure 18) had been replaced by a single main-wheel[92] and one small castor under the outer end of the grain platform. The earlier tendency of the cutter to choke had also been much improved by slotting the underside of the fingers to permit the chaff to escape, and a hinged frame had been adopted. This last feature made cleaning and sharpening the cutters much easier, and also facilitated the fitting of a different cutter for mowing grass or clover. The whole machine was now stronger, lighter, more manoeuvrable, and hence better suited to small wheatfields such as those commonly found in Britain. (See Figure 19.)

Considerable numbers of Hussey reapers were sold in both Britain and America during the 1850s, but the absence of a reel proved a decided disadvantage in tangled or blowing crops. Ultimately, the machines of fellow-American Cyrus H. McCormick were seen as more attractive by the wheat farmers of both countries in terms of both performance and cost, and thereafter Hussey's machines were marketed principally as grass-mowers.

McCormick's Reaper

As noted at the beginning of this paper, some American writers on the history of agricultural machinery have implied (and in some cases, insisted) that the world's first successful reaping machine was invented in 1831 by Cyrus Hall McCormick (1809–84) of Virginia.[93] From the foregoing it is abundantly clear that this was not the case, although McCormick's machine may well have been the earliest American reaper satisfactorily to gather a wheat crop. But that notwithstanding, it is certainly true that McCormick's reaper was more influential than any other in establishing mechanical harvesting in both Europe and America during the second half of the nineteenth century.

C. H. McCormick was introduced to the problems associated with reaper design during childhood by his father, Robert McCormick, who was a wheat farmer at Walnut Grove in Virginia. Robert strove unsuccessfully for more than twenty years to develop a satisfactory harvest-machine for wheat. His first attempt (between 1809 and 1816) ended in failure to achieve even efficient cutting, and had to be abandoned. Experiments with alternative schemes over the next fifteen years proved equally disappointing. In early 1831 Robert produced a machine which cut reasonably well, but unfortunately it left the severed stalks in a tangled mass on the ground.[94] At this point he apparently became thoroughly disillusioned and decided to give up his efforts altogether.

By now Cyrus McCormick was twenty-two years of age, and of course also very well acquainted with the problems which had frustrated his father for so long. He is known to have assisted with his father's last project, and doubtless he shared in the disappointment when it failed. Immediately afterwards, however, Cyrus set out to test a design of his own which he had been contemplating for some time. Construction was begun in May 1831 and the new machine was ready for trial by early July.[95]

Certain surviving records suggest that this first effort by Cyrus was 'a small experimental [machine] . . . mainly designed to test his theory of cutting', and that it was initially tried on a small patch of over-ripe wheat on the family farm.[96] (This had been deliberately left standing at the harvest just completed for such a purpose.) After carrying out one or two tests Cyrus became convinced that:

> he had discovered a practical cutting principle . . . [*He then*] had a locally famous blacksmith make him a new blade with a sickled or serrated edge. Devoting more thought to the equally important consideration of delivering the severed grain in a form fit for binding, he remodeled the divider and added a reel. Thus improved, the machine cut about six acres of oats on the neighbouring farm of John Steele in late July 1831.[97]

According to later testimony by Cyrus McCormick himself, this trial on Steele's farm was carried out on crops of both wheat and oats, and was 'witnessed by many persons'.[98]

No drawing of this first reaper is known to exist, and indeed no authentic diagram of any version of the machine appeared before 1834. However, what is claimed to be a faithful restoration of the original design of 1831 has been published in recent years, and is reproduced in Figure 20. The first detailed description and diagram of the machine (prepared by McCormick himself and published in a New York journal in late 1834)[99] was of the improved version as it stood at that time, but some details of the 1831 original have been extracted from sundry early documents, still extant, as follows:

> This [first] trial machine had a straight, smooth-edged knife,[100] vibrated by a crank and gearing from the main wheel; side draft with the main wheel directly behind the horse, a crude divider, wire or wooden fingers, and a platform of thin plank supported by a small wheel or a slide . . . The main driving wheel was of wood and was two feet in diameter . . . Each finger was double, that is, curved over the edge of the knife, which vibrated in the groove thus formed.[101]

The late July version, which was the first McCormick reaper seen in public, incorporated a number of changes. In a letter written long afterwards McCormick referred to these as follows:

> [the] first rude trial convinced me that a gathering frame for bringing up the grain to the cutting apparatus was necessary . . . I immediately had the additions made to the machine, viz., the gathering reel, with a divider for separating the grain to be cut, from that to be passed by the machine, fixing at the same time a slide instead of a wheel under the grain side of the machine; and adding a serrated cutting blade instead of one with a smooth edge, with improved fingers or guards (over the blade for supporting the grain at the edge while being cut off) at short intervals from each other.

Figure 20. A reproduction of the reaper demonstrated by Cyrus H. McCormick in late July 1831 (from W. T. Hutchinson, *Cyrus Hall McCormick* (1930), facing p. 80).

Figure 21. A McCormick Self-Rake Reaper of 1864 (from Cyrus H. McCormick III, *The Century of the Reaper* (1931), p. 64 and reproduced here by permission of the Houghton Mifflin Company. Copyright, 1931, by Cyrus McCormick).

Instead of having the gear wheels of this machine behind the main driving wheel, they were placed forward of it but outside of the 'driver'.[102]

No provision was made for automatically laying the stalks in a tidy swathe; they were simply swept backwards onto the platform by the action of the reel and the forward motion of the machine. From there they were raked away to the side in bundles of about sheaf-size by a man walking behind or beside the machine.

In 1832 McCormick used his reaper to harvest about fifty acres on the family property at Walnut Grove, as well as several crops belonging to other farmers at Lexington, a few miles distant. One or two further public demonstrations were also given. The following year the machine successfully cut his father's entire crop and again several others on neighbouring properties. By 1834 McCormick had completed a second machine, which included an adjustment for the height of the platform and the cutting-knife. It was also considerably larger than the first, requiring two horses to draw it. Two men now travelled with it: a driver riding one of the horses as before, and a raker standing on the moving platform instead of walking behind. (Cf. Hussey's reaper earlier.)

At this stage McCormick still regarded his reaper as experimental and insufficiently developed to be manufactured for sale. In April 1834, however, he learned of Hussey's patent of the previous December from a description published in a technical magazine.[103] Somewhat dismayed, McCormick at once applied for a patent to protect his own design, and this was granted without difficulty on 21 June 1834. The award was by no means inappropriate, since the two machines differed sufficiently to warrant separate and valid patents for each. Unfortunately, at about the same time as he lodged his application with the patent authorities, McCormick over-reacted in his disappointment by issuing a public warning to Hussey against adopting any of the design features the former considered to be his own.[104] Hussey made no immediate response, but the episode marked the beginning of a series of disputes in the courts as well as in the press which continued until Hussey's death in 1860.

In a centenary volume published in Boston in 1931, McCormick's grandson (Cyrus Hall McCormick III) contemptuously dismissed all earlier claimants to the invention of the reaping machine, and at the same time asserted that:

> no modern grain-cutting machine can suffice without the elements around which Cyrus Hall McCormick's reaper was organized. These essential principles were seven:
>
> The straight reciprocating knife, whereby the standing grain would be attacked by lateral motion as well as by the forward movement of the machine . . .
>
> The fingers or guards for the knife, which supported the grain at the moment of cutting.
>
> The reel, which gathered the grain in front of the reaper and held the heads in place as the fingers held the stalks.
>
> The platform, on which the severed grain might fall to be raked away in a swath.
>
> The main wheel, directly behind the horse, which carried the machine and operated gears to actuate the moving parts.

The principle of cutting to one side of the line of draft, which permitted the horse to walk on the stubble while the cutter bar worked in the standing grain.

The divider at the outer end of the cutter bar, to divide the standing grain from that which was to be reaped.[105]

In fact, only one of these so-called 'essential principles' (the single main wheel) was novel at the time McCormick's machine was introduced. Reciprocating cutters (though admittedly not 'straight' in shape) had been featured more than two decades earlier by Robert Salmon of Woburn (1808?), and again in the machines of Henry Ogle (1822) and Patrick Bell (1828). Fingers of one kind or another had been proposed by several earlier inventors, while the reel had also appeared previously in the designs of Donald Cumming (1811), Ogle, and Bell. The side draught arrangement had been proposed by Gladstone as early as 1806, and the crop-divider by the extraordinary actor James Dobbs in 1815. This latter, incidentally, had also been featured in the American header design of William Manning in 1831, as had also the vibrating cutter already mentioned.

Whether these seven attributes were 'essential' as claimed is very questionable indeed, even if, as is assumed by the younger McCormick, only cutting machines are to be considered. Hussey's reaper, for example, possessed only four of them and yet enjoyed quite a moderate success. But at the same time it is certainly the case that the arrangement of the various elements in McCormick's design was admirably suited to the task of cutting wheat, and with continuous refinement of detail over the years it became progressively more so.

Between 1834 and 1839 McCormick put the development of his reaper aside to concentrate on farming and to take up an entirely new venture in partnership with his father. This last was based on the smelting of iron, for which a good demand and high prices applied at the time. Unfortunately for the McCormicks and others the market price for iron collapsed in 1839; the business failed disastrously and the McCormicks became bankrupt. To meet the demands of creditors, half their land had to be sold and the remainder heavily mortgaged. As a result Cyrus turned once more to the development of his reaper, with a view to making machines for sale as quickly as possible and (hopefully) restoring his family's fortunes.

In 1840 two machines were sold locally, but unfortunately neither of them worked particularly well.[106] This led to further experimentation during 1841 and an important modification to the cutting-knife. The serrations on the cutting edges, which previously had all been angled the same way, were now arranged in groups with alternately opposite angles.[107] A marked improvement in cutting efficiency was the result. Seven machines were sold in 1842, twenty-nine in 1843, and fifty in 1844. The following year McCormick was awarded a second patent, and licensed the manufacture of his reaper to the New York firm of Seymour and Morgan. Sales continued to grow at a steady rate, but McCormick had by now realized that his manufacturing operation was wrongly located. Economic logic dictated that the machines ought to be produced in the region where they would eventually be used, i.e. in the mid-west wheat belt. Accordingly, he terminated the New York agreement as soon as possible and moved to Chicago, where he set up his own factory in 1847. The remarkable progress of this venture, and its subsequent amalgamation with other firms to form the present International Harvester Co. is of considerable general interest, but not relevant here.[108]

By the time McCormick's reapers began to make inroads into the British market (in the early 1850s) they had been developed into relatively refined and reliable machines. By 1862 a special mechanism to deliver the cut corn in bundles of sheaf-size had been added, thus further reducing the time and labour required for gathering and binding prior to stooking. But quite apart from such refinements as this self-rake attachment (see Figure 21), McCormick's reapers did well in sales for more basic reasons; in service they proved to be more durable, lighter in draught, and to need less maintenance than those marketed by rival makers. In addition, the obtuse-angled and serrated cutters were more effective in their actual cutting than those used on either the Crosskill-Bell or Hussey machines. They also usually required less frequent sharpening, and were less inclined to choke in heavy crops. As a consequence, reaping machines manufactured at C. H. McCormick's Chicago works achieved an enormous popularity with both American and British grain farmers. Eventually they triumphed over all other makes on both these major markets and, during the second half of the nineteenth century, they became the most widely used wheat-harvesting machines in the world.

The South Australian Stripper

The kind of circumstances in which the harvest machine now known as the South Australian Stripper (or alternatively, the Ridley Stripper) arose were totally different from those associated with the British and American machines so far discussed. The Stripper was in fact invented and introduced at Adelaide, South Australia, in November 1843 as a direct response to a crisis which had developed in the colony's agriculture. (This, it should be remembered, was a very early stage in South Australia's history; the very first settlers in the region had waded ashore from the migrant ships only seven years previously.) Put at its simplest, the crisis was due to the fact that, at the planting season of the year 1843, South Australian farmers sowed 23,000 acres of wheat — far more than could be harvested the following summer (using traditional hand-reaping methods) by all the available manpower of the colony. Indeed, when the full extent of the sowing became clear, it was seen that more than one-third and perhaps as much as one-half of the crop would have to be left in the fields and wasted.

The thoughts of colonists and colonial officials alike naturally turned to the possible use of machines to save the day, but not with very much confidence, since time was by now very short. As shown in earlier discussion, practical reaping machines working on the cutting principle had been in existence in both Britain and North America for several years, but none of these had yet been manufactured in significant numbers. Certainly none were available for export to Australia. But there was also a further factor which in any case would have made the adoption of any cutting-style machine in southern Australia very unwise and almost inevitably unsuccessful. This was the climate of the region — one of high temperatures and very low rainfall in the summer season — which ensured that wheat crops were always in an extremely dry and brittle state when ready to be harvested. Any attempt to cut the stalks in such circumstances — either by hand or by machine — would result in a heavy loss of grain by natural shedding.

Fortunately, an appropriate local response to the crisis was forthcoming and in time to save the 1843 crop. At the beginning of the harvest season a novel harvest machine was introduced by John Ridley, a migrant flour-miller from the north-east of England. Ridley gave the first public demonstration of his machine on 14 November 1843 on a crop of wheat at Wayville near Adelaide, and it was a triumphant success. Before a large crowd of onlookers the machine worked smoothly and continuously for the whole afternoon, collecting the grain cleanly and at the rate of an acre per hour or better. Only two pairs of horses were required (each being alternately worked and rested for an hour), and two men, one to drive the machine and the other to help with unloading and changing the horses. On several subsequent days the pattern was the same, as was afterwards recorded by one of those present at the actual scene:

> success attended the very first trial of it, and during seven days it reaped and thrashed the *seventy acres* of wheat of which the paddock we all went to see was composed.[109]

However, not only was Ridley's machine immediately successful, it was also very unusual in both its appearance and its principle of operation. As shown in Figure 22, it bore perhaps a superficial resemblance to Bell's Scottish Reaper and one or two other early machines in that the horses were yoked to a long pole at the rear, and pushed the machine before them into the crop. The intention in using this arrangement, of course, was to avoid damage to the unharvested part of the grain by trampling. But there the resemblance between Ridley's and the earlier devices ended; Ridley's machine included no cutting apparatus of any kind, and indeed the stalks were left standing in the field after it had passed. It harvested by beating the grain directly from the ears of the standing crop by means of a set of rapidly revolving wooden flails (or beaters) while the stalks were trapped momentarily in a set of projecting

Figure 22. The original form of the Ridley Stripper, introduced in late 1843. (A sketch by 'N.R.F.' in 1845. From Marcus Collisson, *South Australia in 1844–45* (1845)).

spikes (or comb). These spikes could be adjusted for height so that they apprehended the stalks just below the grain-heads, and so provided temporary support while the beaters did their work.

Because of the extreme dryness of South Australian wheat at harvest-time, the grains separated easily from the ears, carrying only a few husks with them into the machine. Thus the operations of reaping and threshing were performed together in the same machine. Only a simple winnowing operation was required to prepare the grain for grinding into flour at the nearest mill. In general, the straw left in the fields was never cut and collected afterwards, since in South Australia it had little value. Usually it was used to graze sheep, or simply burned off and the ashes ploughed back into the soil.

Two Strippers (both built by Ridley) were employed at the 1843–4 harvest and, together with the efforts of men using traditional scythes and hooks, they enabled the whole crop to be saved. During the following year a number of improvements were made to the design which made the machine more manoeuvrable and easier to steer, the most notable being the removal of the rear grinding pole and the attachment of the horses to a front corner of the structure. In other words, the machine was converted to be drawn from the front with side-draught, as in the earlier overseas designs of Gladstone and others.[110]

By the late 1840s more than half of South Australia's continually expanding wheat crop was being harvested each year by an ever-growing number of Strippers manufactured in and around Adelaide. Thereafter, the gathering of grain-crops by hand-methods gradually disappeared from the local scene. By the 1870s the Stripper had been developed into a refined and very efficient machine indeed, although it was still true that no change whatever of a fundamental nature had been made to it. At this stage it was still, as formerly, providing southern Australian farmers with the most rapid and the cheapest grain-harvesting process available anywhere in the world.

By 1880 more than 10,000 Strippers (all locally manufactured) were employed in the colony of South Australia alone to harvest more than one million acres of wheat, and many more were in use in the drier northern regions of the neighbouring colony of Victoria.[111] Indeed, it can fairly be said that in the dry wheat-lands in this and other parts of the Australian continent the Stripper's triumph was absolutely complete. No other style of harvest machine made any impression there throughout the nineteenth century nor in the first decade or so of the twentieth. Then, as elsewhere in the world, however, the much larger power-operated combine harvesters gradually established their dominance in the harvesting of cereals throughout the continent.

Bell, Hussey, McCormick and Ridley — True inventors or mere improvers?

Truly objective criteria for attributes such as practicability or effectiveness are virtually impossible to specify for any class of machinery. Indeed, even a reasonable consensus of opinion on such matters can prove to be far from easy to obtain. In addition, it is not frequently the case that the extent to which a particular design was truly original may be difficult, or perhaps even impossible, to gauge — as any Patent Attorney will testify.

In the case of the reaping machines of Bell, Hussey, and McCormick, for example, it is relatively easy to show that certain mechanical features of each and all of them had already appeared in earlier designs, and sometimes more than once. Also (and perhaps as a consequence of this) their three machines had many points of similarity with each other. Despite these circumstances, however, there is no record of an admission by any one of the three to having drawn upon the experience and/or ideas of either their predecessors or their contemporaries. On the other hand, John Ridley's South Australian Stripper bore almost no resemblance whatever to any of the aforementioned cutting machines, and it found application in an entirely different region of the world where rather special conditions prevailed. It was also introduced rather later than the others. However, by the mid-1840s these four machines — Bell's in Britain, Hussey's and McCormick's in America, and Ridley's in southern Australia — had emerged as those most likely to bring about the general mechanization of wheat-harvesting, and all others faded into obscurity.

Taking Bell's reaper first, it was shown earlier that the reciprocating cutter-bar, the reel, and the travelling canvas or cloth apron had all been used before his time. A glance at Figures 11 and 13 reveals that the same is true of the long rear pole for pushing the machine in front of the horses. In addition, it might perhaps be thought significant that the earlier machines on which these features had appeared had all risen in close proximity to the area where Bell lived. Bell nevertheless insisted that his 1827(?) design was entirely his own:

> it has been sometimes insinuated that I borrowed my ideas from another, or that I copied from a machine long ago defunct. I deny the truth of all such insinuations. With the exception of Mr. Smith's machine,[112] I had not seen or heard of any other design of the kind . . . It was after my own invention was exhibited that I acquired any knowledge of the machines that existed before my day.[113]

Apart from the similar mechanical features and geographical proximity of the earlier designs already mentioned, there is nothing to suggest that the above assertion by Bell was other than strictly truthful. As was noted earlier, Bell took out no patent for his own design, and so he had nothing to gain financially from misrepresentation. In any case he was a minister of religion and reportedly respected throughout his life as a sincere, conscientious man. That such a man could be a liar and a thief of achievements rightfully belonging to others seems, at least on the face of things, very unlikely indeed.

On the other side of the coin, however, it is clear from Bell's own writings that he suspected — indeed believed — that the American machines of Hussey and McCormick had been derived at least in part from his own.[114] He pointed out correctly that detailed descriptions and diagrams of his reaper had been circulated freely in the U.S. in well-known journals and encyclopediae from 1828 onwards.[115] He also drew attention (again correctly) to the fact that four Bell-pattern reapers were taken to the eastern U.S.A. in the early 1830s. However, in spite of these circumstances neither Hussey nor McCormick ever acknowledged any kind of debt to Bell or his predecessors, or for that matter admitted even to having known of their machines before introducing their own.

Later writers on the subject differed sharply in their opinions. James Slight, for example, suggested that McCormick may have partially copied not only Bell's design but Henry Ogle's as well. But at the same time Slight highlighted his own uncertainty by noting that it was possible that McCormick evolved a similar design independently.[116] Perhaps not unexpectedly McCormick's grandson took an aggressively opposite line:

> He [Cyrus H. McCormick I] had never heard of Pitt's work nearly fifty years before, nor of Bell and Ogle, nor did he know of Manning who had already patented certain of the features he was to discover for himself.[117]

In addition, this author defended his claim for the authenticity of his grandfather's invention on quite separate grounds, as follows:

> Nevertheless, even if there had been no originality in any of the seven features, his reaper was still a true invention. It is a well settled rule of patent law that an invention need actually be no more than a new combination of known features to produce a novel and useful result and be a true discovery.[118]

Doubts of a different kind were voiced by Leander McCormick, younger brother of Cyrus and a sometime executive in Cyrus' manufacturing organisation. According to him, most if not all the principal design ideas for the successful reaper of 1831 were supplied by the father, Robert McCormick, and Cyrus dishonestly took the credit.[119] This was later denied by other members of the family, including the mother (Mrs. Mary Ann McCormick) and another brother (William S. McCormick). Both testified separately and under oath that Cyrus was the true inventor.[120] In view of this and Leander's known personal antagonism towards Cyrus, it would seem likely that the charge was false and motivated by spite.

Whether 'a new combination of known features to produce a novel and useful result' really is sufficient to constitute an invention (as proposed by McCormick's grandson) is arguable. It is true that many patents have been awarded on grounds of this sort, but equally true that many have been overturned later when challenged in courts of law. In McCormick's case the novelty of his single large ground-wheel might just make the difference in his favour.

On the question of whether McCormick deliberately adopted features from earlier designs without acknowledgement, it must be conceded that to date there is no firm evidence to show that he did. Hence, the circumstantial evidence notwithstanding, McCormick must for the present at least have the benefit of any doubt, as well as the credit due to an innovator of considerable talent. Perhaps the most that could be said on the other side is that it might be prudent to keep an open mind until further information is forthcoming. However, what is beyond question is that, at least in later life, Cyrus H. McCormick was an industrial organizer and businessman of genius. In the long run, this was probably far more important in establishing the supremacy of his reaper (in the Northern Hemisphere) than the originality, or perhaps even the merits, of the machine itself.

In Obed Hussey's case there is likewise no direct evidence that he knew of any of the earlier reaper designs before introducing his own in 1833. Indeed, for some time he was apparently also unaware of the existence of McCormick's machine. Hussey's generous acknowledgement of McCormick's priority, when he eventually learned of the latter's public demonstration given at Walnut Grove in 1831,[121] suggests that he was not the kind of man to accept any credit which was not rightly due to him. In this context, note also that Hussey stubbornly refused to add a reel to his reaper to aid its cutting, simply because he had not thought of it himself. Had he done so, his competitive position in the reaper market would undoubtedly have been much improved. Even McCormick's grandson conceded that Hussey was 'an honest old warrior',[122] and all the evidence so far available tends to support that assessment of him. Hence, he too must have the benefit of any doubt as things now stand.

In the case of John Ridley and his Stripper of 1843, the situation is again different. Here there remains a long-standing dispute between Ridley and another South Australian pioneer, John Wraithall Bull, over rights to the invention of the machine's operating principle (a dispute which is too complex to pursue here), but there has never been any suggestion that it owed anything at all to the earlier developments of either the British or the Americans. Ridley himself always claimed[123] that his only source of inspiration was the ancient Roman vallus, a brief description of which he had noticed in J.C. Loudon's *Encyclopedia of Agriculture*[124] some time previously. From this, he said, he had obtained the idea of trapping the ears of wheat in a vallus-like comb. The further idea of adding a beating apparatus above such a comb to knock the grains from the trapped ears seems to have been entirely his own.

Of course, Ridley's standing as an inventor depends in considerable measure upon whether he himself conceived the stripping principle, or whether (as claimed by J.W. Bull and his supporters) Ridley appropriated the idea from Bull, and merely succeeded in applying it at a working level. As I have demonstrated elsewhere[125] the evidence reveals upon close examination that in fact Ridley and Bull must have arrived at the same idea at about the same time and quite independently of each other. The difference between the two is that Ridley was then able to proceed to a successful practical application of the stripping idea, whereas Bull never progressed beyond displaying a small model at a public meeting in Adelaide. Thus it is seen that Ridley possessed an inventive talent of a rare, complete type — one in which a capacity for imaginative innovation was allied to a sure feel for practical feasibility. Certainly it is true that his design was much more radical than any of the three considered earlier, and indeed it was a more immediately successful one as well.

Conclusion: The influence of the Bell, Hussey, McCormick, and Ridley Machines upon the progress of general mechanization in harvesting

As already seen, the introduction of mechanical harvesting on a widespread basis was delayed in both Britain and America for approximately two decades after suitable machines had made their appearance. In Britain, Patrick Bell's Scottish Reaper could have replaced hand harvesting with the scythe-and-cradle from 1828 onwards, as is evidenced by the machine's continuing usefulness on the

Bell family farms in Scotland over very many seasons. In fact, however, it did not, and the traditional hand-methods of reaping grain persisted.[126] Bell's reaper, along with those of Smith of Deanston, Henry Ogle, and others, was ignored by farmers and potential manufacturers alike until the early 1850s.

There can be no doubt that if the social conditions of the time had been more propitious, farmers would have found the Bell reaper to be a cost-saving as well as a labour-saving device.[127] Calculations made by Professor James Hendrick[128] suggest that for a crop of moderate size (say 50 or 60 acres) the cost of reaping using Bell's machine in 1832 would have been four shillings per acre or less. The cost of reaping by hand at that time was about eleven to thirteen shillings per acre.

In America the situation during the late 1830s and early 1840s was somewhat different. Hussey's reapers were available for purchase from 1834 onwards, and those of McCormick from about 1840, but the farmers remained unconvinced of their supposed benefits. In point of fact, the farmers' judgements were correct; neither type was particularly reliable until around 1842. Concerning his own machine, McCormick himself afterwards admitted as much when he wrote:

> No machines were sold until 1840 and . . . were not of much practical value until my second patent, 1845.[129]

When McCormick's reaper was functioning properly, however, its advantages were easily seen. It would then cut from ten to fifteen acres in a day with an attendant labour force of only two operators and five binders.[130] To accomplish the same work manually, between six and ten scythe-hands were needed for the cutting and again five or so men to bind the cut stalks into sheaves.

Hussey's reaper was simpler in construction than McCormick's and thus probably the more reliable of the two in the early years. In addition, Hussey's cutting apparatus was clearly superior to McCormick's until the early 1840s at the least. It was superior also to the clipping cutter employed by Patrick Bell, and it is noteworthy that the latter spoke openly of Hussey's cutter design and of his ability as a mechanician generally with 'unfeigned respect'.[131]

The Hussey and McCormick machines finally began to displace hand-reaping on American wheat farms after 1849. The opportunity came when large numbers of agricultural workers east of the Rocky Mountains abandoned their rural employment to seek fortunes on the newly discovered gold-fields in California. Within a few years these two machines captured the major part of the British market as well, as already noted. In a sense, this last was more than a little ironical, since British inventors had been the true pioneers of mechanical harvester design. In addition, the successful American machines were not only similar in many important respects to those introduced by Ogle and Bell during the 1820s, but they also operated on virtually identical principles.

The Ridley Stripper on the other hand never found favour in either England or America, and indeed its use was always restricted to the south of the Australian continent. There, however, it consistently performed superbly, while other machines of the cutting variety (like the McCormick and the Hussey) were entirely unsuitable, and indeed were never introduced in substantial numbers. Cutting the crop and stooking to dry the grain was found to be generally unnecessary in Australia, since the climate itself ensured that the grain was well dried by the time it was fully ripe.

But this is not to say that the Australian stripping approach is the only one suited to regions where the summer season is hot and dry. In California, for example, grain harvesting by the heading technique has always been preferred, i.e. the stalks were cut just below the ears and only the grain-heads were collected. The straw left behind in the field was later fed to stock or alternatively used as a mulch for the soil. Combined header-thresher machines were also developed quite early in California, and they proved very successful indeed.[132] Once again, ordinary reaping machines of the cutting variety found almost no use there whatever.

In the dry wheat lands of southern Australia, the Ridley Stripper reigned supreme as a harvest machine from soon after its introduction in 1843 until virtually the end of the nineteenth century. Like the Hussey and McCormick machines in America, the Stripper had maximum impact upon harvesting practice in its region after a severe labour shortage developed on Australian wheat farms in the early 1850s. As in the American case, this Australian labour deficit was also caused by a rush of agricultural workers to newly discovered gold-fields — this time at Ballarat and Bendigo in the colony of Victoria. But as is now well known, the Australian wheat industry survived this upheaval and indeed underwent a remarkable expansion during the latter half of the nineteenth century. Huge quantities of Australian grain were supplied to the markets of Britain and western Europe, and the Australian colonies grew rich on the proceeds. The steady profitability of the wheat trade was due to a number of factors, and certainly not least among these was the widespread use of the Ridley Stripper, which afforded Australian grain farmers the lowest harvesting costs then available anywhere in the world.

However, if we return finally to a general consideration of the pattern of development which occurred in the mechanization of cereal harvesting around the world, three features are evident:

1. Its adoption was essentially a mid-nineteenth-century phenomenon.

2. A number of early reaper designs which were themselves unsuccessful nevertheless included important features which reappeared in later, successful machines.

3. Three separate approaches to cereal harvesting were developed in particular regions (viz. cutting, heading, and stripping), and all three succeeded in appropriate conditions.

Cyrus H. McCormick's American reaper was unquestionably the central machine in the change-over to mechanical harvesting in the cool, moist wheat-growing regions of the Northern Hemisphere. The designs of both Bell and Hussey faded from the picture from about the late 1850s, and never reappeared. Machines of the header type proved effective in the drier climate of California, but this approach to harvesting was restricted to that relatively small region until approximately the second decade of the twentieth century. In Australia, mechanization was accomplished almost solely by the Ridley Stripper, and proceeded quite independently of harvest-machine developments in other parts of the world.

Notes

1. See Cyrus Hall McCormick III, *The Century of the Reaper* (1931).

2. *Nature*, Vol. 138 (1936), p. 1088.

3. *Nature*, Vol. 139 (1937), pp. 197–8.

4. A few proposals for reapers were also patented in the U.S.A. during this period, but the first American design of any real promise appears to have been that of William Manning in 1831.

5. A full list of Fussell's publications appears in a booklet entitled *G.E. Fussell : a Bibliography of his Writings on Agricultural History*, Museum of English Rural Life, University of Reading, 1967.

6. Patent No. 13,962 (London), 9 February 1852, 'Cutting and Reaping Machines'.

7. G.E. Fussell, *The Farmer's Tools: 1500–1900* (1952), p. 130.

8. *Country Life*, Vol. 122, No. 3,173 (7 November 1957), p. 981.

9. Reproduced from W. Harcus (ed.), *South Australia, its History, Resources and Productions* (1876).

10. K.D. White, *Roman Farming* (1970), note 46 for chapter 7, on p. 486.

11. *Ibid*.

12. See A.B. Lees, *Farming Machinery* (1951), p. 102, and also Plates 42 and 43.

13. K.D. White, *Agricultural Implements of the Roman World* (1967), p. 173.

14. M. Partridge, *Farm Tools Through the Ages* (1973), p. 132.

15. A controversy over who invented the fundamental principle of the Stripper — Ridley or Bull — has been carried on in an unfortunately acrimonious fashion from 1845 until the present day.

16. See, for example, G.L. Sutton, 'The Invention of the Stripper', *Journal of the Department of Agriculture of Western Australia*, Vol. 14, 2nd series, No. 3 (September 1937), pp.193–247. However, Sutton's conclusions on the Ridley–Bull controversy are by no means universally accepted.

17. See A.R. Callaghan and A.J. Millington, *The Wheat Industry in Australia* (1956), chapter 21.

18. See G.B. Wilkinson, *South Australia: its Advantages and Resources* (1848). In a following book, *The Working Man's Handbook to South Australia* (1849), Wilkinson compounded his error by claiming that the machine placed the grain in bags as well (see p. 40).

19. See Jacob Wilson, 'Reaping Machines', *Transactions of the Highland and Agricultural Society of Scotland* (January 1864), pp. 123–49.

20. This has sometimes been a source of confusion. In the U.S.A. and in Australia, for example, the term 'corn' is an alternative name for maize rather than for wheat.

21. See L.A. Moritz, *Grain-Mills and Flour in Classical Antiquity* (1958), especially chapter 2.

22. See C. Singer, E.J. Holmyard, and A.R. Hall (eds.), *A History of Technology*, Vol. 2 (1956), particularly chapters 4 and 17. Note, for example, the group of sixteen large water-powered mills for grinding corn constructed at Arles in Roman-occupied France

between AD 308 and 316. The use of water-power for driving mills afterwards spread throughout Europe and the British Isles over some eight or so centuries.

23. For information on the early wind-driven mills of Britain, see the numerous articles by Rex Wailes in the *Transactions of the Newcomen Society* between 1929 and the mid-1970s. Wailes' book, *The Windmills of England* (1954), is a summary of some of these.

24. 1786 is the year quoted in most references, e.g. in Merrill Denison, *Harvest Triumphant* (1948), p. 12. However, in *Man the Maker* (1958), p. 229, R.J. Forbes states that the threshing machine was 'perfected' by Meikle in 1788. A description of Meikle's machine (with diagrams) is available in J.C. Loudon, *An Encyclopedia of Agriculture* (1825), pp. 403–6, and elsewhere. (See also note 26.)

25. The rapid introduction of winnowing machines in Scotland in the early nineteenth century is noted, for example, in an article by 'Mr. Sullivan' included in *The Farmer's Friend* (editor not identified), (1847), p. 61ff.

26. An excellent treatment of the evolution of threshing and winnowing machinery is provided in Robert Ritchie, *The Farm Engineer: a Treatise on Barn Machinery* (1849).

27. See H.A. Beecham and J. Higgs, *The Story of Farm Tools*, 2nd edition (1961), pp. 28–9.

28. *Ibid.*, pp. 30–1.

29. See A.R. Callaghan and A.J. Millington, *The Wheat Industry in Australia* (1956), p.333.

30. Merrill Denison, *Harvest Triumphant* (1948), pp. 12–13.

31. Pliny (the Elder), *Historia Naturalis*, Book 18. See vol. 5 of the translation by H. Rackham (London: William Heinemann, 1950), p. 375.

32. For a translation of the pamphlet and a detailed commentary, see E.A. Thompson and B. Flower, *A Roman Reformer and Inventor* (1952). Thompson and Flower suggest that the author of the pamphlet concealed his identity deliberately, in order to safeguard himself from reprisals from the Roman authorities over his criticisms of public and military administration.

33. Palladius, *De Re Rustica*. Translation from J.C. Loudon, *An Encyclopedia of Agriculture* (1825), p. 26.

34. From Slicher van Bath's article 'The Influence of Economic Conditions on the Development of Agricultural Tools and Machines in History', included in J.L. Meij, *Mechanization in Agriculture* (1960).

35. Trevieren is the name of the Gallic tribe which settled in the Montauban region before the birth of Christ, and built a fortress there. This tablet was apparently part of a decoration on a wall of the fortress.

36. G.E. Fussell, *Farming Technique from Prehistoric to Modern Times* (1966).

37. K.D. White, *Agricultural Implements of the Roman World* (1967), chapter 10.

38. For conjectural illustrations of the two machines, see *ibid.*, p. 157.

39. A misleading description, since the machine was in fact a crude header.

40. From an (unsigned) article entitled 'The Ancient Mechanical Harvester of Gaul', *Agricultural Engineering* (October 1973), p. 50.

41. See E. Thompson and B. Flower, as note 32, footnote to p. 80.

42. Cf. the establishment of a set of sixteen large water-powered flour-mills at Barbegal, near Arles, during the peropd AD 308–16 to grind flour for the same armies. See C. Singer *et al.*, *A History of Technology*, Vol. 2 (1956), pp. 598–9.

43. K.D. White, *Roman Farming* (1970), p. 183.

44. H.A. Beecham and J. Higgs, as note 27, p. 30.

45. Lynn White, Jnr., 'What Accelerated Technological Progress in the Western Middle Ages?', in A.C. Crombie (ed.), *Scientific Change* (1963), pp. 272–91 (p. 274).

46. Jacob Wilson, 'Reaping Machines', *Transactions of the Highland and Agricultural Society of Scotland* (January 1864), pp. 123–49 (pp. 123–4).

47. M. Denison, as note 30, p. 13.

48. Patent No. 2,324 (London), 3 August 1799. Incidentally, in F.L. Wheelhouse, *Digging Stick to Rotary Hoe* (1966), Boyce's first name is incorrectly given as James. The correct name of Joseph is shown on the patent specification itself.

49. Patent No. 2,404 (London), 14 June 1800.

50. See Jacob Wilson, as note 46, p. 124.

51. Patent No. 2,873 (London), 20 September 1805.

52. In his caption for this illustration, Fussell claims that it was obtained from *Encyclopaedia Britannica*, 1797 edition. In fact, no such illustration appears in that edition, and indeed one would not expect to find it there. As Fussell himself indicates (on pp. 117–8), this machine and that of Robert Salmon (which is supposed to have been taken from the same source) both appear to have been introduced in the first decade of the nineteenth century.

53. G.E. Fussell, *The Farmer's Tools: 1500–1900* (1952), p. 118. Fussell's source appears to have been an unsigned article entitled 'Patent Report on Reaping Machines and their Inventors', *Quarterly Journal of Agriculture* (1855), pp. 611–23.

54. H.A. Beecham and J. Higgs, as note 27, p. 31.

55. Figure 10 is formed of these five figures from Brown's work printed together.

56. Patent No. 3,468 (London), 16 November 1811.

57. Dobbs patented his design—Patent No. 3,844 (London), 21 March 1815.

58. G.E. Fussell, *The Farmer's Tools: 1500–1900* (1952), p. 119.

59. James Slight, 'Report on Reaping Machines', *Transactions of the Highland and Agricultural Society of Scotland* (January 1852), pp. 183–98, especially p. 189.

60. *Ibid.*, pp. 189–90.

61. Jacob Wilson, as note 46, p. 125.

62. This diagram and an accompanying description were published in the *Mechanic's Magazine* (12 November 1825), pp. 49–50.

63. M. Denison, as note 30, p.15.

64. Estimates vary considerably. One to two acres per day for a man using scythe-and-cradle is the rate suggested by R.L. Ardrey, 'The Harvesting Machine Industry', *Scientific American Supplement*, Vol. 54, No. 1407 (20 December 1902), pp. 22544–7. F.L. Wheelhouse, in her book *Digging Stick to Rotary Hoe* (1966), p. 50. estimates about one acre in an afternoon (equivalent to approx. two acres per day), while H.A. Beecham and J. Higgs (note 27) give a figure of 2.3 acres per day for a man with a scythe.

65. Jacob Wilson, as note 46, p. 125.

66. James Slight, as note 59, p. 186.

67. Jacob Wilson, as note 46, p. 125.

68. The display card alongside the exhibit at the Science Museum states explicitly that it was 'invented in 1826.'

69. Patrick Bell, 'Some Account of Bell's Reaping Machine', *Quarterly Journal of Agriculture* (January 1854), pp. 185–204, especially pp. 187–9.

70. Some time before the machine was taken to the Science Museum in 1868, the cutting mechanism had been modified. However, part of the original cutter was also located, and this is now displayed with the machine.

71. Patrick Bell, as note 69, p. 186.

72. *Ibid.*, p. 191.

73. *Ibid.*

74. *Ibid.*, p. 192.

75. Because Patrick Bell's father had grown old and retired from farming.

76. G.E. Fussell, *The Farmer's Tools: 1500–1900* (1952), p. 124.

77. *Ibid.*

78. See Philip Pusey, 'Report to H.R.H. the President of the Commission for the Exhibition of the Works of Industry of All Nations', *Journal of the Royal Agricultural Society of England*, Vol. 12 (1851), pp. 587–648. See particularly the sub-section entitled 'Harvesting Implements—1. Reaping Machines'.

79. Patrick Bell, as note 69, p. 193.

80. Details of some of these contests are to be found in Jacob Wilson's 1864 article (note 46), in E. Stabler, *Overlooked Pages of Reaper History* (1854), and elsewhere.

81. See Baron Ernle (R.E. Prothero), *English Farming Past and Present*, 6th edition (1961), pp. 316–7.

82. According to Jacob Wilson (note 46), p. 126, an allegedly effective reaper was also in existence in Russia by this time.

83. This was by no means the first American machine to cut in this fashion however.

The earliest American patent for a machine of the header type appears to have been issued to one Samuel Lane of Hallowell, Maine, in 1828. (See W.T. Hutchinson, *Cyrus Hall McCormick*, Vol. 1 (1930), p. 157).

84. *Ibid.*, p. 62. The crop-divider had, of course, appeared earlier in England. (Cf. Gladstone's reaper of 1806, etc.)

85. Reynold M. Wik, 'Some Interpretations of the Mechanization of Agriculture in the Far West' [of America], *Agricultural History*, Vol. 49 (1975), p. 79.

86. Records show that he was born in Maine in 1792. He died in 1860 as a result of being struck by a train.

87. See W.T. Hutchinson, as note 83, pp. 459–60.

88. Jacob Wilson, as note 46, p. 126.

89. See W.T. Hutchinson, as note 83, p. 162.

90. Jacob Wilson, as note 46, p. 127.

91. See W.T. Hutchinson, as note 83, p. 172.

92. Adapted from C.H. McCormick's design; see following section.

93. See, for example, Cyrus H. McCormick III, *The Century of the Reaper* (1931); also W.T. Hutchinson, as note 83.

94. See Cyrus Hall McCormick III, as note 93, p. 8.

95. See W.T. Hutchinson, as note 84, pp. 79–80. According to Hutchinson, Robert McCormick did 'some of the rough carpentering work . . . under Cyrus' direction' (p. 81).

96. *Ibid.*, pp. 82–3.

97. *Ibid.*, p. 83.

98, McCormick's own statement, published in the (New York) *Mechanic's Magazine* (May 1834), and quoted in W.T. Hutchinson, as note 83, pp. 83–4.

99. This first appeared in the (New York) *Mechanic's Magazine* (11 October 1834), and has since been reproduced in W.D. Rasmussen, *Readings in the History of American Agriculture* (1960).

100. As noted earlier, this was replaced by a new serrated blade immediately after the first tests carried out (privately) on the McCormick family farm in July 1831, and before the public demonstration given on Steele's property later the same month.

101. W.T. Hutchinson, as note 83, pp. 81, 82, and 82n.

102. Cyrus H. McCormick to H.C. Parsons, 27 September 1881; quoted in W.T. Hutchinson, as note 83, footnote to p. 82.

103. (New York) *Mechanic's Magzine* (April 1834). See also W.T. Hutchinson, as note 83, p. 91.

104. This was in the form of a published letter. It is quoted in W.T. Hutchinson, as note 83, pp. 91–2.

105. Cyrus H. McCormick III, *The Century of the Reaper* (1931), pp. 17–18.

106. See W.T. Hutchinson, as note 83, p. 181.

107. *Ibid.*, p. 183.

108. For an account of the development of the agricultural-implement manufacturing industry in America up to and including the formation of the International Harvester Co., see R.L. Ardrey's 1902 article (note 64).

109. James Milne, *Romance of a Pro-Consul* (1899), p. 64.

110. These improvements and other aspects of the Stripper will be discussed in detail in a separate paper, which it is hoped may be included in a future volume of *History of Technology*.

111. For wheat acreages by years in the various Australian colonies, see W. Harcus, as note 9, table facing p. 394.

112. Smith of Deanston (see the section on the first modern attempts to devise mechanical reapers).

113. Patrick Bell, as note 69, p. 186.

114. *Ibid.*, pp. 193–200.

115. For example, a full description of Bell's reaper (with diagrams) was provided in the 1831 edition of J.C. Loudon's *Encyclopedia of Agriculture*, pp. 422–7. Descriptions of the designs of Smith of Deanston and of Henry Ogle were also included.

116. See James Slight, as note 59, pp. 196–7.

117. Cyrus H. McCormick III, as note 105, pp. 9–10.

118. *Ibid.*, p. 18.

119. These views were published by Leander J. McCormick in a volume entitled *The Memorial of Robert McCormick* (1885). For a discussion of the controversy, see W.T. Hutchinson, *Cyrus Hall McCormick* (1968), chapter 5.

120. See W.T. Hutchinson, as note 119, p. 111.

121. *Ibid.*, pp. 64–5.

122. Cyrus H. McCormick III, as note 105, p. 25.

123. See his published letter on the subject in the *South Australian Register* (6 May 1886). The letter is reproduced in the Appendix to G.L. Sutton's article (note 16), and also in Annie E. Ridley, *A. Backward Glance* (1904).

124. This description appeared in both the 1825 and the 1831 editions.

125. In my Doctoral dissertation entitled 'The Contributions of John Ridley to Australian Technology and to the Early Progress of South Australia'. (A copy is filed in the Baillieu Library at the University of Melbourne, Australia.)

126. As late as September 1843 at the Northumberland Agricultural Society's annual show held at Hexham, a certain Mr. William Temple reportedly exhibited 'a new corn scythe . . . [which] will do as much as four hooks used in the ordinary way'. This report was reprinted (presumably from a Northumbrian newspaper) in the Adelaide *Observer* (10 February 1844), p. 2, cols. 1, 2.

127. However, W.T. Hutchinson (note 83) insists that Bell's machine *was not* prevented from succeeding in Britain between 1828 and 1851 by the labour situation, as is popularly supposed (see pp. 65–66), but his argument is unconvincing and his view remains very much a minority one.

128. See J. Hendrick, 'Patrick Bell and the Centenary of the Reaping Machine', *Transactions of the Highland and Agricultural Society of Scotland*, 5th series, Vol. 40 (1928), pp. 51–69.

129. C.H. McCormick, quoted by Philip Pusey, 'On the Progress of Agricultural Knowledge during the Past Eight Years', *Journal of the Royal Agricultural Society of England* (1850); and in G.E. Fussell, *The Farmer's Tools: 1500–1900* (1952), p. 127.

130. See W.T. Hutchinson, as note 83, p. 180.

131. Patrick Bell, 'Some Account of Bell's Reaping Machine', *Quarterly Journal of Agriculture* (January 1854), p. 198.

132. See Roy Bainer, 'Science and Technology in Western Agriculture', and Reynold M. Wik, 'Some Interpretations of the Mechanization of Agriculture in the Far West', both in *Agricultural History*, Vol. 49, No. 1 (January 1975).

The Sector and Chain:
An Historical Enquiry

G. HOLLISTER-SHORT

The idea that machines, like genera and species in the natural order, might be subject to processes of change of a quasi-evolutionary nature is scarcely novel and yet very little has been done to throw light on how, when one has divested the question of metaphorical distortion, the process might work itself out, what its significant features might be and its modalities.[1] This is perhaps the more surprising because such an approach, when addressed to the problem of the evolution of tool forms, showed long ago what a rich field was waiting to be exploited.[2] As with tools, so with machines: might not one expect to achieve comparable insights into the nature of invention and the importance of gestalt forms?[3] Such considerations seemed sufficient reason to pursue the history of a simple yet elegant device, the sector and chain, which was to be exploited in a steadily widening series of applications because of the valuable property it possessed of permitting two different sorts of motions, circular and straight line, to be coupled in a continuously connected kinematic arrangement. An analogous device, the sector and rack, will also be touched upon but no more than lightly since only a separate treatment would permit its own interesting history to be dealt with adequately. It might be as well, at this point, to mention also that the modern term sector and chain is rather anachronistic. Before the period of the Newcomen engine it would be more accurate to talk of sector and rope: certainly there are no sectors and chains in Leonardo's manuscripts.

It would, however, be of slight interest to consider any transmitting mechanism apart from the motor machine and the mechanized tool associated with it, and in fact the device has a wider interest. It was always associated with a machine type that was, no matter what modifications might be made to it (for the sake of having it run by an inanimate prime mover, for instance) always of the balance beam or swape type: and this is as true of the latest manifestation which this enquiry will record, of *c.* 1775, as it is of the earliest. Also as constant a feature, at least until the eighteenth century, was its employment as a component of pumping machinery. Beyond dispute the kinematic properties of the sector and chain were valuable, and one has only to look at Francesco di Giorgio Martini's earlier approach to the problem[4] to appreciate its economy: but in any case its survival in the inventories of engineers through three centuries is sufficient evidence of that. (See Table, which gives a chronology of sector and rope/chain devices.) An interesting question, but one too vast to address here, would be to consider why this means of converting oscillating motion into rectilinear motion should have been preferred to other methods.

The sector and chain device was invented by Leonardo da Vinci, or so much may be said provisionally since I have not been able to find any earlier evidence for the combination. It would appear in addition to be one of the earliest of his inventions that we know of since the device appears in MS.B, of 1488–9, the oldest of the codices which can be dated, although examples do occur in some of the later MSS as well, one of which, f76v in MS.L., it will be instructive to look at in due course. It is scarcely possible, in the nature of the case, to date drawings in the *Codex Atlanticus* although there are two sketches there which must be considered and which in a sense, as I hope to show, might be considered as logically prior to part of MS.B. even though this probably cannot be established with certainty in any chronological sense. The group of Leonardo's drawings that I wish to consider are: MS.B. f6v, f20r, f53v, f54r, f70r; MS.L. f76v; *C.A.* f16v.c., f57v.a., f392r.b.; and *Codex Forster*, III, f45r.

In MS.B. f54r the drawing of a pendulum operated rocking beam, 'dalzare acqua', '(figure 1) shows the sector and chain idea very clearly: as each end of the beam (which is really nothing more than two sectors put apex to apex) lifts, it raises the piston of a suction pump which, once it has reached the maximum of its upward excursion, falls back under gravity. This may be taken as the classic expression of this sort of pump, and seems to have provided the model which an engineer such as Thomas Newcomen, for example, was to follow some two hundred years after Leonardo's initial statement of it. An earlier drawing (f20r in the same MS) of a pump, 'da fare montare acqua', (figure 2) shows the form the pump had to take when the pistons were of a Ctesibian valveless type where water was to be forced to a height rather than lifted from a depth. In this situation both excursions of each of the pistons needed to be performed under power, and the attractiveness of this model was to make itself felt as late as 1779 for it was with such racks and sectors that Matthew Wasborough was to experiment and which Watt himself was to employ until 1786.[5] But Leonardo had also sketched in MS.B. f53v a machine with a single rack and sector (figure 3) and written beside it: 'E possi fare con corda come con rota dentata' (And one can make it with ropes as well as with a toothed wheel). Although it is to anticipate somewhat, it may be noted here that the idea of doubling the rope in the sector and rope/chain device so as to render it capable of pulling in both directions was also to commend itself to engineers after Leonardo's time. Curiously enough, the sources which disclose the continuing employment of the sector and chain reveal also, in two instances, the use of the double rope device.[6]

But where had the sector and rope/chain come from? It seems to me that if one is to make any attempt to explain the genesis of MS.B. f54r

TABLE

Chronology of sector and rope/chain devices

Double sectors

1488	L. da Vinci	pendulum pump
1582	P. Morris	component of pumping engine
1617	J. de Strada	pendulum pump
1652	W. Schildknecht	component of pumping engine
1659	C. Huygens	chronometer balance
1662	J. Böckler	pendulum pump
1675	C. Huygens	chronometer balance
1693	C. Huygens	chronometer balance
?1710	J. Schmidt	*libra communis*
1712	T. Newcomen	Dudley Castle engine
1726	J. Leupold	meteorological instruments
1734	J. Höll	*Hebelmaschine*
1756	Anon.	pendulum pump
1760	J. Sisson	balance barometer

Single sectors

1488	L. da Vinci	pile driver
1502	L. da Vinci	excavating device
?1615	V. Scamozzi	adjustable water wheel
c.1640	G. Desargues	pumping engine
1669	C. Perrault	balance bob
1693	J. Hadley	adjustable water wheel
1694	P. de La Hire	pumping engine
1701	G. Sorocold	adjustable wheel
c.1708	Anon.	vat tilting device
1711	M. Höll	T. bobs
1724	H. Sully	chronometer balance
1730	G. Gerves	pumping machine
1733	A. Barnes	T. bobs
c.1750	Anon.	chimney cranes
1771	W. Henry	register for stoves
c.1775	Anon.	candle dipping

Double ropes in place of rack and sector

1488	L. da Vinci	pumping machine
1582	P. Morris	pumping machine
1652	W. Schildknecht	pumping machine
c.1700	Anon.	ships' tillers
1712	T. Newcomen	plug rod drive
1721–5	R. Newsham	pumping machine
1767	J. Smeaton	pumping engine

(figure 1) one has to look at the repertory of machine types with which Leonardo had presumably become familiar in his early years as an engineer. One of these was certainly the manganon, a type of trebuchet such as one finds in the *Liber Tertius de Ingeneis . . .* of Taccola composed about fifty years before Leonardo's time.[7] In such an engine the axle to which the trebuchet arm is fixed was loaded with weights in such a way that at rest the sling arm was in the vertical position and would obviously need to be wound down before the machine could be loaded and fired. All this Taccola shows us clearly enough. In Leonardo's hands, however, the machine was undergoing development. Evidently he was dissatisfied with it. In MS.B. (figure 4) he provided the throwing arm with a wheel, and although the latter is shown without ropes, these were evidently intended, as is indicated by the pegs protruding from the rim of the wheel. The throwing arm and its weight were to be drawn down into the firing position by tractive effort exerted on the wheel although whether this was to be applied from beneath or from the side is difficult to determine. At least, however, the separate elements of Taccola's machine (the rollers and the winch) had been rationalized (or so it would appear) in Leonardo's wheel manganon. But once modification of it had begun the process could be carried further. Consider the *modus operandi* of Leonardo's wheel trebuchet in the *Codex Atlanticus*. f57v.a. (figure 5). The latter is shown as a complete wheel but with a lower half of solid construction (the previously separate weight). A cord is secured round the rim of the wheel, and its 'free' end, hanging down, is attached to a block which in the drawing is 'frozen' just as it is on the point of disappearing through the trap beneath it. The trebuchet arm has been jerked violently upward, and its sling cords, pulled out straight behind it, are on the point of discharging the missile. Leonardo does not show how the machine was to be got ready for firing in the first place although it seems reasonable, in view of his other drawings, to suppose that a rope hanging down on the opposite side of the wheel from the weight would be pulled upon by the crew. But did the attached weight, evidently added to increase the firing power of the machine, now make such a solid and bottom heavy wheel unnecessary? Perhaps in fact its construction, once the weight had been hung from it in this way, needed to be made as light as possible since the importance of the wheel would now reside hardly at all in its weight but rather in its function as a *rope-carrier*. It can scarcely have escaped Leonard's notice that movement is greatest at the periphery of a wheel and that the largest possible fraction of the counterweight part of the apparatus should be concentrated where the pull, resulting from the sudden release of the block, would be vertically downwards. It is at this point that it becomes important to examine another drawing in the *Codex Atlanticus*, the trebuchet in f16v.c. (figure 6). If we pursue the idea of the wheel as a rope-carrier, it is evident that one quarter of its circumference (the quarter missing in f16v.c.) can perform no function at all in this respect so that now there would be every reason to remove it.[8] In the three quarter wheel

Figure 1. Pendulum pump of *c.* 1488–9 after Leonardo, MS.B. f54r.
Figure 2. Force pump of 1488–9 after Leonardo, MS.B. f20r.

Figure 3. Force pump of 1488–9 after Leonardo, MS.B. f53v.
Figure 4. Wheel manganon of 1488–9 after Leonardo, MS.B. f6v.

there was indeed some positive advantage in doing this since it ensured that the attached weight, able to be raised higher, should fall from a greater height. This extra fall was important because Leonardo now planned that the throwing arm should describe an arc about twice as great as that travelled by the arm of the previous machine. The positions of the weight and the sling make it obvious that the arm must move clockwise through something like 180° as the weight falls. In *C.A.* f16v.c. (figure 6)

Figure 5. Wheel manganon from Leonardo, *Codex Atlanticus*, f57v.a.

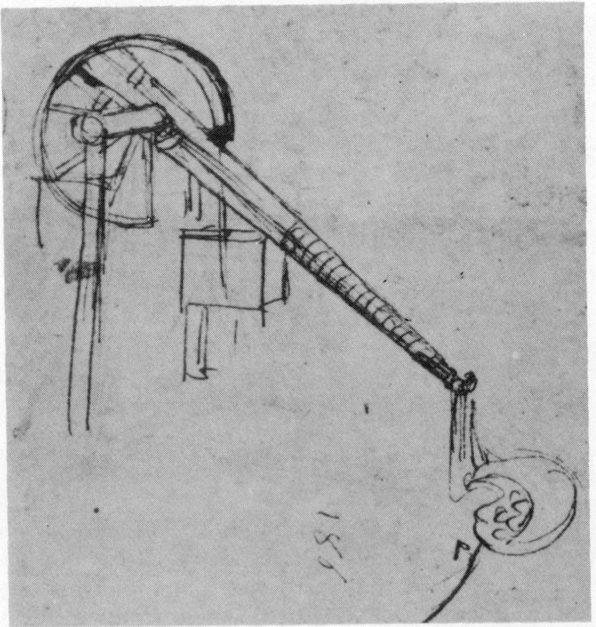

Figure 6. Wheel manganon from Leonardo, *Codex Atlanticus*, f16v.c.

the machine is not shown discharged but is either in the cocked position and ready for firing or, more probably, at an instant after release since the counterweight could not be raised very much more before it and the advancing lower edge of the sector began to foul each other. The rope on the pulling side of the wheel would seem to have been previously drawn down from beneath. Nevertheless what we actually see, since Leonardo chose to draw it in this particular position, is a machine that in some respects is like f57v.a. (figure 5) turned upside down. Even if the new type of trebuchet was never constructed it would not require prolonged inspection of the sketch (rather indeed a spontaneous flash or moment of eureka) to see that the reversal might go further and that the alternate pull on either side of the wheel from beneath by means of cords in order to swing the trebuchet arm might with advantage be put into reverse: why should not the arm be swung to produce an alternating upward pull on the ropes and consequently upon whatever was attached to their ends? In this general post of all the elements why should not the throwing arm turn into a pendulum, whose movement would cause the wheel to oscillate? (Its bag of stones to provide momentum is already there.) This movement would cause the cords hanging from its sides to be pulled alternately upwards and thus work, as it might be, suction pumps to which they were attached. Another quarter of the circle becomes redundant, producing the symmetrical shape of MS.B. f54r (figure 1) while at the same time the protean swape re-emerges from the form in which it had, so to speak, lain concealed in Taccola's and Leonardo's trebuchets. If Truesdell is right in thinking that the essential quality of Leonardo's greatness lay in his unfaltering power of observation and grasp of kinematic relationships, then it might well be the case that changing the characteristics of one machine permitted him to see, unwilled, the form of another emerging when all its essential elements were under his hand.[9] If this is the case then MS.B. f54r (figure 1) follows and so also perhaps do the Madrid codices which according to Reti are a 'systematic treatise of practical kinematics'.[10]

If a new machine had disclosed itself it remained to explore its properties and to see in what situations it might be possible to exploit it. A number of recent studies have rightly emphasized Leonardo's importance as an engineer. It is obvious from the considerable number of sketches to do with the task of pile-driving that Leonardo had given much thought to this basic and constantly recurring task. In MS.B. f70r occurs a proposal, 'da fichare pali a castello' (to drive piles at a castle) (figure 7) involving a sector and chain device that is in every sense intimately related to *C.A.* f16v.c. (figure 6). A crank winds down the erstwhile trebuchet arm in a thoroughly Taccolan way and raises the weight, now become a pile-driving ram, which hangs from a cord fixed to the quarter circle at its further end. On release no dead donkey is hurled over any city wall: the violent movement following release is channelled through the cross head guides directly downwards on to the head of the stake. Another principal

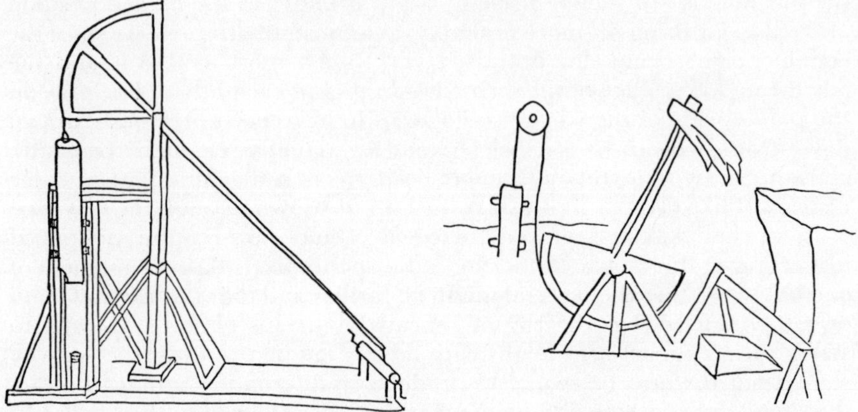

Figure 7. Pile driver of 1488–9 after Leonardo, MS.B. f70r.
Figure 8. Excavating device of 1502–3 after Leonardo, MS.L. f76v.

area of activity lay in hydraulic engineering, and, as Reti has shown, it was here that Leonardo gave great thought to the problem of how to handle the removal of earth, particularly in the cutting of canals, in the most economical way.[11] MS.L. f76v (figure 8) which is directed to this problem probably dates from 1502–3, the period when Leonardo was employed in the Romagna in the service of Cesare Borgia. The excavating device he now sketched is another variation on the sector and chain idea, and comparison with MS.B. f70r (figure 7) is well worth while because like the pile-driver it throws light on the sort of device with which Leonardo probably began his trebuchet experiments, that is to say the bottom-heavy Taccolan type in which the throwing arm is in a vertical position when the machine is discharged or at rest. Compared to the pile-driver the flow of energy through the elements of the excavating machine is exactly reversed. Work is put in at the pile-driver end, now become a device with rings upon which a team of men pull, with the result that the sector rope or chain is pulled violently upwards. The cutting head of the excavating tool swings down and takes a slice from the face of the cutting and tumbles the spoil into the basket placed at its foot. The cut completed, the cutting head returns by itself to its vertical at-rest position at the same time hauling the pulling weight back up in readiness for the next heave upon it. When Reti quotes Leonardo's words describing the ideal situation,'the soil removed would jump by itself quickly on the instrument that will transport it',[12] one feels that Leonardo would have been entitled to add the words 'and the tool which performs the labour will swing back by itself ready to work again'. Such seems to have been the way Leonardo came to recognize this new gestalt form and proceeded to exploit it. If the first step is difficult, the subsequent ones are easy: 'difficile est invenire, facile autem inventis addere'.

One can only speculate as to the manner in which Leonardo's ideas were disseminated: that they had begun to make their way in some measure even from the earliest period into the inventories of other engineers need not be doubted. Like any other engineer of his time Leonardo learned his business from current practice, from contact with his fellow engineers, and from manuscript literature. The ideas that he gathered from these sources were subjected, as I have tried to show, to vigorous scrutiny and re-formulation. The new ideas and forms that resulted, insofar as they were incorporated into machines actually built, were there for all to see. Even if they were only discussed, they would, in another sense, become public. For a Renaissance engineer could not practise behind closed doors like a Benjamin Huntsman or a Samuel Crompton presenting the world with a finished product whose acquisition even by a purchaser with a professional interest in the mystery would reveal nothing worthwhile about its fabrication. If an engineer were to show his superiority at all then he could not confine himself to the pages of his notebook; he could hardly avoid divulging his arts in some fashion to the interested gaze of his fellows, whose attention would be drawn in proportion to the success which a new device achieved. We do not know, of course, which ideas were exploited and which remained paper schemes, but it is becoming increasingly obvious that even if one were to take an extreme case, that is, to suppose that nothing was constructed at all, we should still be wrong to think that his ideas remained locked up and inaccessible in his notebooks. A number of formal parallels have been shown to exist between Leonardo's sketches and certain of Ramelli's machines, to whose number I hope to add another example presently, which make it appear almost certain that Ramelli had been able to study Leonardo's manuscripts himself. But why should one, in any case, imagine so absurd a situation as that of an engineer who never engineered anything? Probably whatever he did was incorporated almost immediately in the inventories of other engineers just as he had freely absorbed other people's ideas himself. In fact, as far as the question of diffusion is concerned, Reti has put the matter as plainly as possible, 'Leonardo did not work in a vacuum but in close contact with fellow engineers, artisans and helpers'.[13]

From the last quarter of the sixteenth century evidence survives which shows, I think, not only that at least three of the group of machine sketches which I have listed had survived very much as Leonardo had formulated them and were in circulation, in several senses of the word, in France, Germany and England, but also, for such is the nature of the evidence, that far from being content to repeat the Leonardian schemes, a new generation of engineers was seeking to adapt them to more exacting situations in order to get more work out of them. Why they were doing this, or rather why they continued so much more vigorously and successfully in these endeavours than other societies is not a question that has received any very satisfactory answer but concretely it could not but result in a continuous evolution of machine types. Pressed to the limit some device would cease to be serviceable save in a new formulation involving familiar elements, or unless out of some creative metamorphosis, such as would seem to have produced the sector and chain, a new superior form became available. Nevertheless it should not be thought that any machine or device, however simple, would need necessarily to disappear because a superior device had in some situations superseded it. An ecological niche for it would continue to exist wherever an unexacting situation continued to exist: the swapes of the slate quarries of Angers, the source of Joachim du Bellay's 'ardoise fine', co-exist in the pages of the *Encyclopédie* with the fire engine at the Bois de Bossu. Indeed, this is to put the matter almost the wrong way round: there must at that time have been hundreds, if not thousands, of swapes for every atmospheric steam-engine.

On 23 February 1633 a fire on London Bridge, above the first arch on the north side, destroyed some of the houses and laid open to view the pumping engine which had been installed in the arch over fifty years previously. It was observed by John Bate who published an account and a drawing of the machine in 1635 in the second edition of his *Mysteries of Nature and Art*. After mentioning that it pumped water into a tower whence distribution pipes supplied an area two miles across, he tells us how he got to see it, 'Which engine I circumspecitvely viewed as I accidentally passed by immediately after the late fire that was upon the bridge Anno 1633, and the device seeming very good when I came home I drew a modell thereof and have here represented it to the view' (Figure 9). Either in 1575 or 1577 Sir Christopher Hatton was petitioning the Crown that a patent for an engine 'to draw and raise up water higher then nature of yt selfe only serveth' should be granted to his follower, Peter Morris.[14] The patent, granted on 24 January 1578, spoke of this 'new kynde and manner of engynes . . . not nowe or heretofore as we are informed . . . made, practized or used by any other within this our realm of England . . . wythin memorie of any man'. But construction did not then begin. Financial difficulties of various sorts ensued even after Morris had secured a contract in 1580 for supplying Thames water to Leadenhall, and the engine only finally began to pump late in December 1582. It is most probable that this engine was mounted on a boat since Bate shows no arrangement for adjusting the wheel to the state of the tide. It was Rhys Jenkins's belief that certain tubes, nearly six inches in diameter, recovered from the site when excavation for a new bridge were begun in 1828, were the machine's pump barrels.[15] Fynes Moryson, writing in 1591, may well provide the answer if his remarks apply to Morris's work. In that year Moryson was in Dresden and noted the 'water mills swimming upon boats and removed from place to place the like whereof was since made at London by a Dutchman but became unprofitable by the ebbing and flowing'.[16]

However, it is the actual constitution of the engine which is of interest here, and several of its features demand inspection. Its mode of operation

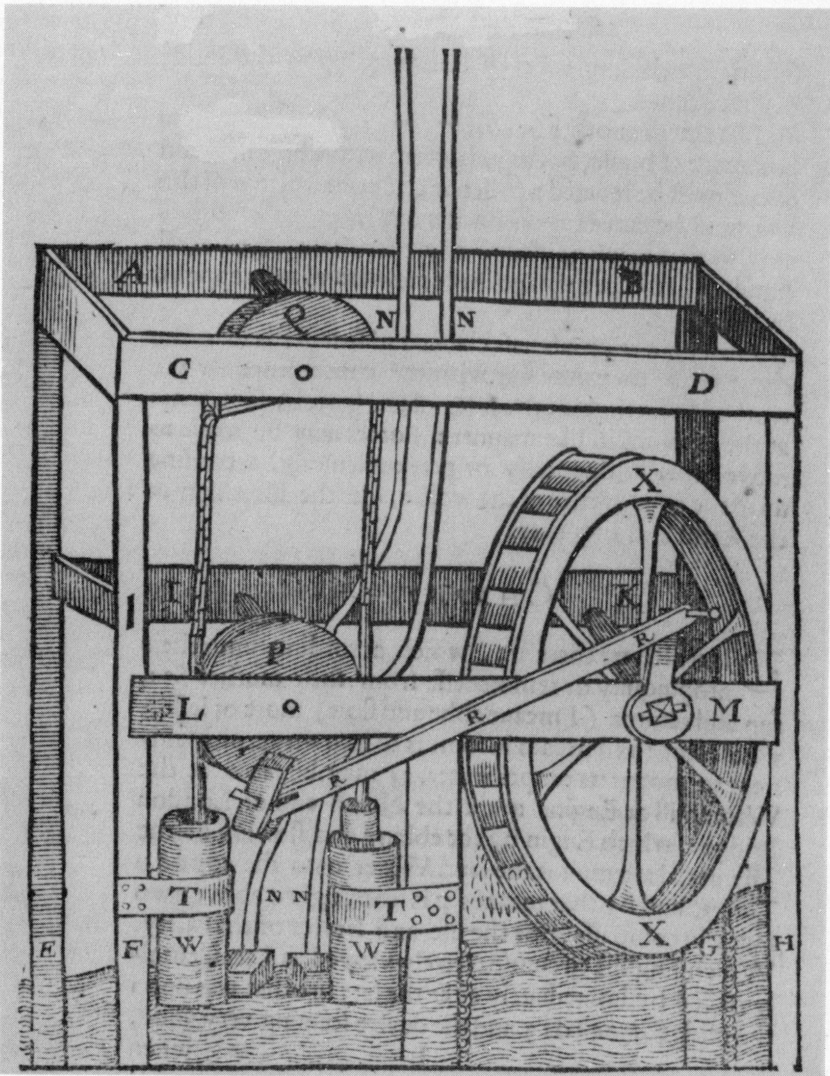

Figure 9. Morris's London Bridge machine of 1582 from J. Bate, *Mysteries of Nature and Art*, London 1635.

may be stated quite simply. The tidal flow of the Thames served to move the water wheel (X) in either direction which motion the connecting rod (R) communicated to the small wheel (P). P was required only to oscillate, and the necessary intermittent motion is likely to have been achieved in the following manner. If, as appears to be the case, the water-wheel (X) is intended to be shown revolving in a clockwise direction (and how else should the water be spilling from it?) then the connecting-rod (R), secured by a pin, between the fork pieces fastened to the bottom of the small wheel (P) will pull it round in an anti-clockwise direction until the fork has described an arc of about 90° from its starting point. By that time the forcer (*embolus*) in the pump nearest the wheel (X) will have been drawn up to its fullest extent in preparation for its forcing stroke which, of course, the other plunger will have just completed. The small wheel will then be forced backwards in a clockwise direction to repeat the sequence of events in reverse order. The function of sector Q and its chains above the small wheel was to locate the pump rods and keep them perpendicular. I hope it will not seem like labouring the obvious to say that Morris's arrangements, once he or some precursor took the decision to dispense with the pendulum of the pump in MS.B. f54r (Figure 1), follow as a logical consequence. In order to get the work done by a rather more powerful prime mover than, for instance, the poor fellow to be seen in Figure 10 operating Strada's version of Leonardo's machine,[17] it was necessary to find a combination of mechanisms which would yield the two basic motions of the manually worked device: reciprocatory (the thrust of the worker on the pendulum) and oscillatory (the sectors moving about their centre). The connecting-rod and small wheel reproduce these motions so exactly that it is difficult not to see the simple progenitor haunting this altered form like a ghost. Morris's particular choice of means to transform the pendulum pump into a water-wheel driven affair was, of course, far from being the only one by which this might have been achieved. Philip Skippon sketched a machine at Augsburg in 1663 (one of a number, he says) in which mutilated gears on the axle of the water-wheel worked directly on the racks with which the pump rods were fitted.[18] Leonardo himself (in *Codex Forster*, III, f54r) had figured, long before, an inordinately complicated arrangement by means of which a pendulum pump could be worked by animals. It has to be said, however, that the scheme was his only project for such a conversion so that it would appear to be the case that the adaptation of the pump to animal or water-power was to be the achievement of others working after Leonardo's time rather than of Leonardo himself. A further complication faced by Morris was the need to force water to a height of over one hundred feet. Previous descriptions of this engine have, however, faltered womewhat when the moment came to decide how the oscillating motion of P was transferred to the force pumps disposed either side of it. It is possible, for instance, that partial sets of gear-teeth, although they are not indicated, engaged with pins on the pump rods. On balance, this seems a little unlikely since Bate does not show the morticed ends one would expect to see on P if it had half-lanterns. Indeed, the simpler layout of Skippon's Augsburg machine would seem to be a form in which they would more naturally find a place. It seems unlikely also that the pump rods could have had teeth or pegs since Bate tells us that they were only two inches thick. One is forced therefore to envisage, although there is no

Figure 10. Pendulum pump from J. de Strada, *Künstliche Abriss*, Frankfurt 1617, Vol. 1, fig. 43.

hint dropped about it in either text or drawing, that an arrangement of chains, one running under P and one above it, linked P to the tops and bottoms of the piston rods rather like a double sector and chain, a counterpart to the single one which appears marked as Q above P. Leonardo had noted besides MS.B. f53v that ropes could do the work of toothed sectors (Figure 3) so that Morris's machine appears on this argument still more Leonardian in inspiration. The rack and sector idea of MS.B. f20r (Figure 2) was still available, as I hope to show, but more probably it was easier and cheaper to adapt the double sector and chain idea to wheel P although in the nature of the case no part of P could be cut away. Bate talks of 'two chains of iron which must be linked straight up to the two ends of an iron band that must compasse the circumference of the uppermost wheel'. Their function was to assist in pulling up the piston rods alternately while the oscillations of wheel P, acting in an opposed sense, would cause the chains running over and under the wheel each to pull down one piston and force up another alternately.[19]

A drawing of a pump (Figure 11) published in 1652 by Wendelin Schildknecht[20] the constructional details of which, he explains, had been shown by him by a boat-builder (*Bootsknecht*) he had met in Holland worked exactly on this double-pull chain principle although since it was a light, portable machine, ropes were used, and the two wheels, P and Q of Morris' machine, are nowhere visible since they are subsumed in the rocking roller (bell-crank) arrangement at the top of the assembly. This roller, which was pulled in see-saw fashion by two men, imparted the necessary reciprocating motion to the pumps. Each pump was attached at top and bottom to a single rope which passed in a loop around the roller. Since the machine delivered power on both excursions of the pistons it is no surprise to find Schildknecht commenting that it could be made to work both suction and force pumps. His charming simile is worth quoting: 'Es kan beydes in ein Truck und Zugwerck transferiret werdē/als ein Strohut/der für die Sonn und Regē gut' (It can be adapted for both force and suction work just as a straw hat is good for sunshine and shower alike).[21] It might seem that a drawback of MS.B. f54r (Figure 1), in view of all this, would lie in its dependence on gravity for its downward excursion but this would be to miss the attraction that its elegance and economy had for engineers after Leonardo's time. Only a slight loading of the piston would presumably be necessary to iron out any hesitancy in return, even in a small light version of the machine, and the fact that the most distinguished engineers in Europe and England were later to exploit the idea would seem to indicate that the machine was still then in use. In the Newcomen engine of 1712, of course, or Mathias Höll's *Stangenkünste* of 1711, the weight of the pump rods was sufficient to ensure their prompt return after each lift. (Höll is mentioned again later on.) Where, as at London Bridge, force pumps were necessary it is interesting to see the Leonardian model preserved by duplicating its chain action.

If the Leonardian background to Morris's work be accepted it still remains to draw on further late sixteenth century evidence to show that yet another of Leonardo's schemes was in circulation. Agostino Ramelli's machine number 32 (Figure 12) has some affinity with Morris's enginé as will now appear but derives its morphology and its kinematics primarily from *Codex Atlanticus* f392r.b. (Figure 13) rather than from MS.B. f54r

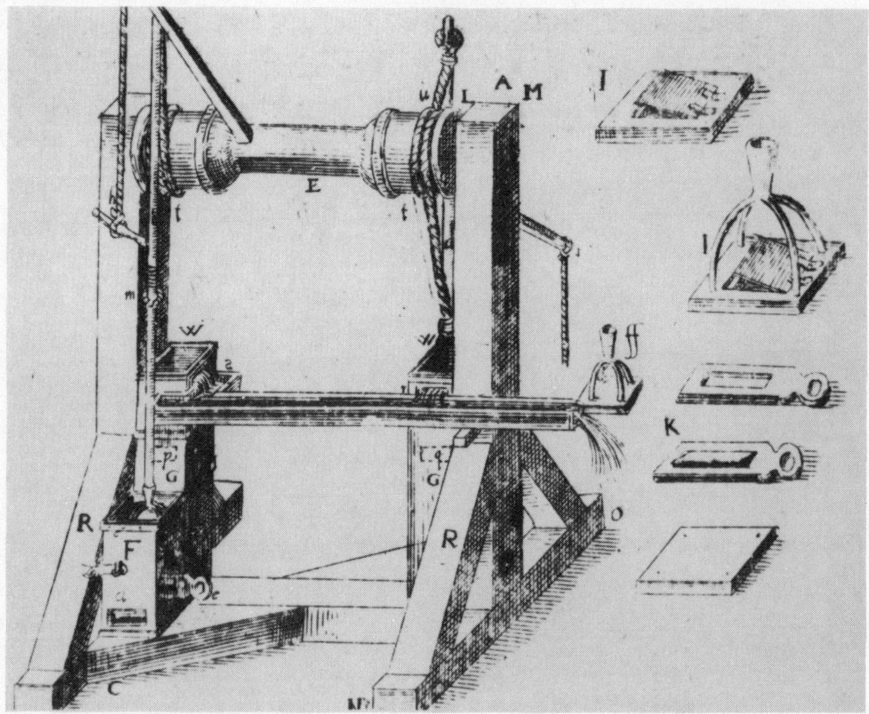

Figure 11. Pump with double-rope drive, from W. Schildknecht, *Harmonia in Fortaliciis*, Stettin 1652, Part III, p. 116, Plate Z, No. 1.

(Figure 1).[22] When the handle of machine number 32 is turned, the half lantern marked H begins to engage the toothed sector S, forcing it downwards and so causing its other end to lift the piston in suction pump D. The chain to which the top of the descending sector is fastened is pulled down and this serves to raise the further sector to which its other end is fastened. This sector is shown on the point of disengaging from its half lantern. As the further sector is raised, so its other end is forced down and depresses the piston in P, the water beneath the piston being forced through T into the receiving basin G. On the face of it Morris's machine appears more economical in its use of elements, since all its motions are in the same plane, while Ramelli's requires a change of direction of motion through 90°. Morris's lower wheel in effect does the work of Ramelli's two sectors but any lengthy comparison of the two machines would scarcely be profitable since they are, so to speak, only distantly related. The true parent of Ramelli's machine is, I suggest, the pump in *C.A.* f392r.b. (Figure 13), and what Ramelli has done in effect has been to get rid of the pendulum drive, using for this purpose the very obvious solution of crank and axle. Although only a lad turns the wheel the arrangement discloses

FIGVRE XXXII.

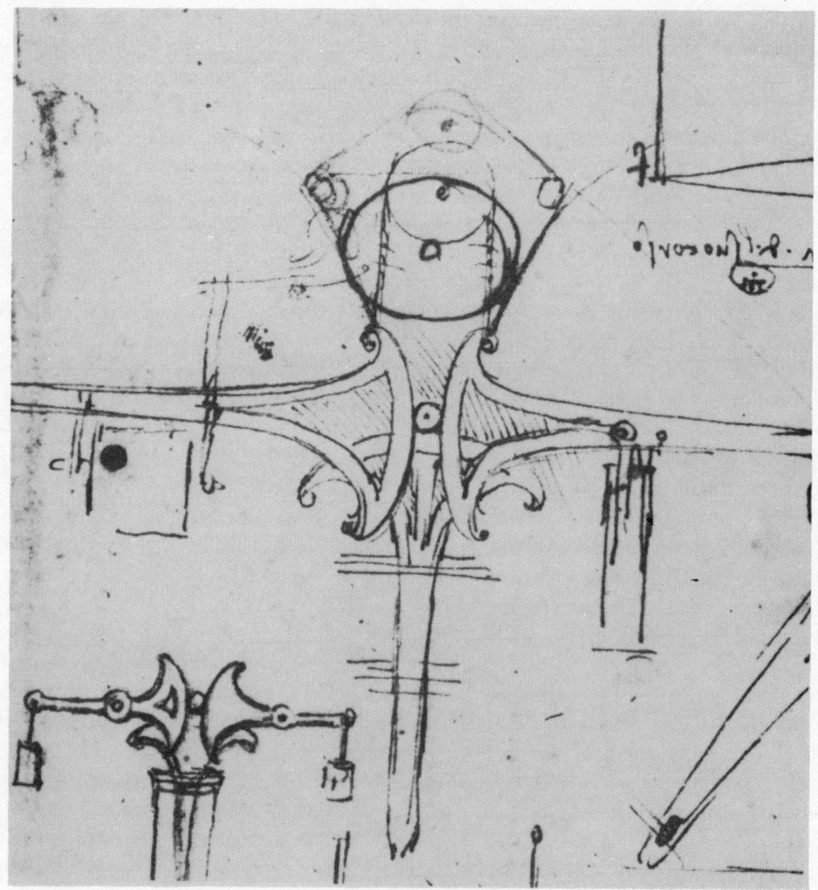

Figure 13. Linkage over pulley, from Leonardo, *Codex Atlanticus*, f392r.b.

that a more powerful prime-mover could be employed at will. But as is often the way, one thing leads to another. Unless the operator is continually to change the direction of his cranking (and presumably it was this to-ing and fro-ing which was objected to in the pendulum drive and impelled Ramelli to abandon it in the first place) it was impossible to retain either Leonardo's friction roller, the end of which appears in the drawing at the top of the pendulum, or the two swape beams facing and linked to each other by a chain passing over a freely revolving upper wheel. The inelegancies of Ramelli's machine, unless it metamorphose still further, now emerge as an inevitable result of the change. The

---

Figure 12. Suction pumps with chain linkage, from A. Ramelli, *Le Artificiose et Diverse Machine*, Paris 1588, Plate XXXII.

swapes, with two sectors, must now be placed side by side, for only so could two half-lantern wheels be made to work them if the cranking were to be performed unidirectionally, although a bonus was to result from this arrangement in that a common rising pipe was now an easy matter to arrange since the two suction pumps were disposed side by side. Observe, however, another effect of all this on the ropes, or, as here, the chains. They will be pulled awkwardly out of alignment once the machine begins to work. It is little wonder that Ramelli does not show this, or that he leaves the upper wheel hanging freely so that it may turn with the slantwise pulls upon it. Such deformations, although present in Leonardo's machine, would have been less troublesome since they were at least confined to the same plane of motion as the wheel. As a final parallel the close resemblance between the open work form of Ramelli's sectors and their curled loops, 'en forme de l'ancre' as he says in the text, to which the chain/rope is attached, to those of Leonardo might even suggest that Ramelli himself had seen Leonardo's drawing, unless all sectors were by then so shaped.

By the early seventeenth century there is the evidence of Strada, already glanced at, to demonstrate that the pendulum pump formula was still alive but the sort of enlargement of the idea which men like Morris had achieved was far from being played out. In 1652 the already mentioned German engineer Schildknecht, a citizen of Stettin, published a book largely concerned with the techniques of military engineering.[23] However, chapter II of the third part of his book is concerned with pumps of various sorts, and what is even more to the point, and worth pages of *Fraktur*, is that it is embellished with two pages of copper plates. He writes with some wit and lets drop many hints of his travels: this pump was shown him in Holland, that one he saw in Padua when he was travelling in the Venetian *terra firma*; and probably a close reading of his book would reveal a great deal more of his experience and travels. He was a reading man and refers approvingly to several of Ramelli's designs. The second of the copper-plates in chapter II shows a pump (Figure 14) profitable (according to Schildknecht) to a ruler or magistrate who wants a water-pumping appliance which is strong, shows skill in construction and is efficient. The central vertical shaft is driven through its lantern by wheel A which here would be worked by a crank. The endless screw on the shaft transmits this movement to the horizontal shaft with which it engages. This shaft has half-lanterns at each of its ends which engage alternately with the toothed sections of the piston-rods working in the pump barrels B and C. The racks of the pistons are kept in close contact with the lanterns by their upper parts being caused to pass between guide beams provided with anti-friction rollers. The downward power stroke resulting from the meshing of lantern and piston rod serves to pull the other rod up: this is secured by the chains to which the ends of both piston rods are connected being linked up over the top of the sector above them, a feature shared with Morris's machine. The Leonardian Kinematics which the latter so

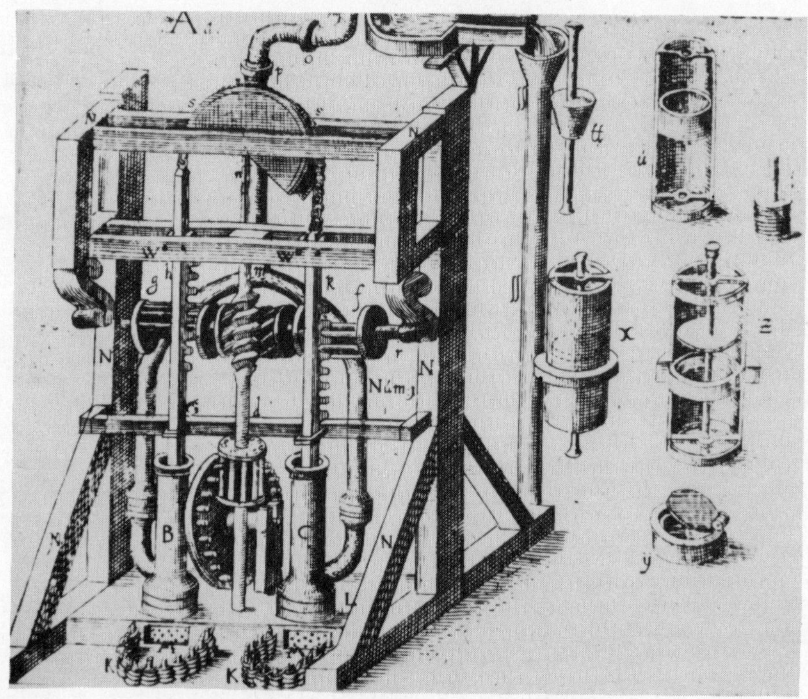

Figure 14. Sector and chain locating device, from W. Schildknecht, *Harmonia in Fortaliciis*, Stettin 1652, Part III, p. 120, Plate A.

slavishly imitated are here handled differently, though whether with any gain in efficiency may be doubted. It has already been pointed out that Morris's device could easily have been adapted to suction work. Schildknecht also shows us a pair of force-pumps but indicates that they too could easily be adapted. It is also of some interest to note that although his drawing shows the sort of arrangement that would be needed were the prime-mover to be a man, he makes it quite clear that the machine could readily be modified. If there were a river with a strong current available, he says, the crank, pin wheel and lantern on the central drive could be replaced by a rimless, horizontal waterwheel. If, on the other hand, water-power was not to be had, but a large delivery of water was required so that syrens, Neptunes and water-snakes spouting freely might provide a heart-easing spectacle for promenaders in a pleasure garden, then again a horizontal wheel worked by men or animals was the solution. The employment of a horizontal wheel, however powered, to work the pumps is an interesting development. If Morris's scheme was transparently the offspring of its Leonardian parent, it is no longer possible to say as much for Schildknecht's, neither the morphology nor the kinematics of which distinctly recalls those of the prototype to mind.

If the prototype had by this time ceased to exist as a machine in everyday use, the problem presented by the next development to be considered would be simple indeed, for the sort of evolution, or as it may rather appear, saltation, for which Schildknecht's machine is evidence, would have left any engineer wishing to effect improvements with one basic problem only: how best to reconcile the rotary motion of the wheel with the reciprocating motion of the pumps it was to operate.[24] It would, after all, be difficult to suppose that Schildknecht's arrangements represented the last word on the subject. But in fact the prototype continued in existence alongside a variety of machines owing something to it and it might well be that some process of synthesis is what we have now to inspect. At about the very time probably when Schildknecht, home from his travels, was arranging his book a French contemporary, Girard Desargues, engineer and mathematician, was constructing a machine at the château of Beaulieu about twenty miles outside Paris whose elegance and simplicity would make the German pump with its lanterns, racks and endless screw look decidedly mannered. It is most unfortunate that Desargues's career as an engineer is, except in this one instance, completely veiled from us despite the fact that it was probably in this field, rather than as an architect, that he was mostly in demand.[25] It is to a mere accident that we owe the earliest description of this machine. On 22 September 1671 Christiaan Huygens wrote to his brother Lodewijk, mentioning Beaulieu but saying almost nothing about the pumping machinery there.[26] Apparently this letter never arrived, or was somehow delayed, because on 29 October 1671 Christiaan was writing again to Lodewijk and refers to its non-arrival. He proceeded to describe again his trip to Beaulieu and Viry. This time, however, the pumping machinery at the former house is the subject of a sketch and a careful explanatory note: 'Pour des fontaines il n'y en a point, que par les moyens de pompes, qui vont par une belle machine de fabrique de M. des Argues. Un mulet y fait tourner une grande roue, qui par le bas, est taillé on ondes, qui en passant sur un rouleau le font baisser et hausser, et en mesme temps le bras auquel est attaché le piston de la pompe'.[27] Its upkeep will cost very little, he says, because not a single toothed wheel is needed, and he goes on to suggest that if Monsieur le Prince (the future William III of England) had not yet built his machine at Honselerdijck he would do well to take that of Beaulieu for model. Huygens, much impressed by the waved wheel, makes it abundantly clear that Desargues, not Römer, was the first to use an epicycloidal wheel. Huygens does not bother with many other details but does indicate that a sector was used on the pump end of the balance beam. Quite when Desargues constructed the machine it is impossible to say. He was in Paris from 1624, and more or less uninterruptedly so from 1630 until shortly after 1648. It seems almost certain that the work belongs to this period. Although he was in Paris again in 1657 and 1658 he had not then long to live.[28] It was only later that an exact idea of its arrangements (Figure 15) emerged from Philippe de La Hire's *Traité des*

Épicycloïdes et de leurs Usages dans les Méchaniques of 1694.[29] It is later still that we learn from Belidor something of its dimensions.[30] In the preface La Hire says, apropos of the usefulness of geometry in mechanics, that because it was his fortune to build a wheel at Beaulieu, 'à la place d'une autre semblable qui y avoit été autrefois construite par M. Desargues et qui étoit entirement ruinée' and because so far as he knew that excellent geometer had only worked out the epicycloidal properties of the wheel mechanically, this was a matter he intended to deal with formally under Proposition IX.[31] Within two years the work had been translated into English and presented unblushingly, as if it were theirs, by Venterus Mandey and James Moxon, although it was tucked away as Book 10, along with other materials in a book tricked out with the catchpenny title, *Mechanick Powers, or the Mysteries of Nature and Art Unvail'd*.[32] La Hire is the earliest writer to comment explicitly on the usefulness of the sector and chain (the quotation is from Mandey's translation): 'Tis easily seen that the chain which is fastened to the portion of the circle serves to raise the pestle always perpendicular which is a very good use in these sorts of pumps; for otherwise if the handle which carries the pestle be only fastened to a leaver movable about an axis as D in this engine, it will happen that the pestle will be drawn sometimes to one side and sometimes to the other, and wear unequaly in the body of the pump in working which will destroy it in a third part of the time, as I have observed in some rencounters'.[33] The engraving of the machine taken by Mandey from La Hire's book along with the text is not as explicit as one would wish about the arrangements of the parts. The revolving wheel or cam-follower which the waved wheel depresses is in fact attached by a bracket to one end of a swape beam centrally pivoted, at whose other end is the sector and chain. Across the waved wheel another cam-follower works in an opposed sense, that is, it will always be at the top of a wave when the other is at the bottom. Belidor shows this much more clearly (Figure 16), giving dimensions and constructional details, as well as showing both bottom waved and top waved wheels.[34]

The enquiry up to this point has been concerned only with the fortunes of certain Leonardian pump forms employing the sector and chain idea from their first appearance in the notebooks. The existence of both in association has so far been traced to the close of the seventeenth century but Leonardo, as we have seen, recognized the value of sector and chain in other applications quite distinct from their use as an element of pumping machines, and there is some reason for supposing that this part of his legacy had survived also and was indeed by the close of the century on the verge of an incredibly rapid extension. Reti has already drawn attention to Leonardo's sketch of a mill-wheel which might be adjusted by means of screws so as to be always in an optimum position in relation to the water which moved it. Ramelli has several designs for such wheels and a brief paragraph in Scamozzi shows quite clearly that these were not paper schemes.[35] The idea had achieved by Scamozzi's time a remarkable

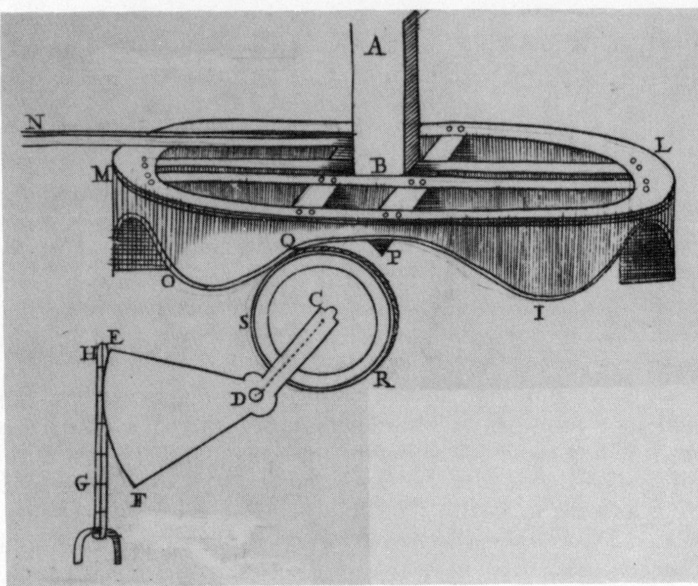

Figure 15. Face cam of the Beaulieu machine, from P. de La Hire, *Traité des Épicycloïdes*, Paris 1694.

geographical extension for he mentions a mill at Noyon in Champagne where the wheel and mill sluices could be lifted more than twenty feet, another at Strasbourg, and others on Lake Lucerne and south of the Alps on a tributary of the Piave. It is clear, however, that by Scamozzi's time a new feature had crept into the design of some of these mills. At a mill on the river Deman near Lucerne he talks of great screws that lift the whole body of the wheel by moving a balance to which the wheel is attached. Scamozzi's word *bilancia*, meaning swape beam, is something new in this context. There is nothing in Leonardo's drawing or Ramelli's designs which could possibly have called for such a word since they notably lack pivoted beams. As is nearly always the case with verbal descriptions, one is left in a state of uncomfortable doubt as to what the thing being described actually looked like. However, what may have been the nature of the arrangement begins to emerge when the first information comes to hand of adjustable wheels in England.

The first definitive evidence is in the patent secured by John Hadley of Worcester, No. 613 of 1693, for 'raising and letting down vertical wheels so as to render them useful at all heights of the water'. Worcester, where Hadley is said by Dickinson to have first employed the device, would certainly be a place where such adjustments would be extremely useful since the river Severn is subject to great and rapidly occurring variations of level. Quite when Hadley began work with another engineer, George Sorocold of Derby, is uncertain, although it seems that Sorocold had used

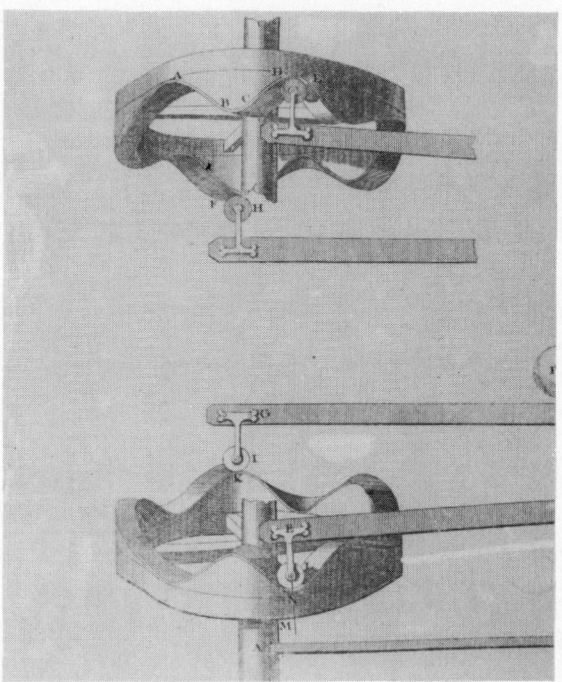

Figure 16. Face cams from B. F. de Belidor, *Architecture Hydraulique*, Vol. 2, Paris 1739, Part I, Ch. 4, Planche VII.

Hadley's device, even before it was patented, for the mill-work he had completed at Derby in 1692. Celia Fiennes, who visited Derby in 1692, and who had always an eye for money-getting schemes, was impressed by 'the rising and falling wheel' at Sorocold's works, and explained, 'at this engine they can grind if it is ever so high a flood which hinders all the others from working . . . they are quite choaked up'.[36] George Sorocold's greatest work was, however, to come in 1701 when the new proprietors of the London Bridge Waterworks commissioned him to build two great wheels in the fourth arch of the bridge. Once again Hadley's patent device was used, and when Henry Beighton drew the machine in 1731 we finally see what it was, or at least the particular form it took in this application. It is only by looking through a small thicket of pipes, beams and connecting rods that one is able to see what was involved. Beighton's description is very precise.[37] There are 'two great levers (L,N) . . . The wheel (G) is, by these levers, made to rise and fall with the Tide'. Then follows a formal exposition of the lettered members (the levers are sixteen feet long) before he concludes: 'One man, with the two windlasses (W), raises or lets down the wheel as there is occasion . . . By means of this machine the Strength of an ordinary man will raise about fifty Ton weight'. But was it all worth while? 'The machine for raising and falling the wheels is very good, though but seldom used as they tell me; for they (the wheels) will go at almost any depth of water, and as the tide turns the wheels go the same way with it.'

Dickinson thought that perhaps Beighton was being told the tale by smooth operators who had found the changing times of tides too monstrously variable to live with and who were quite prepared to let the wheel take its chances. It is also possible that Sorocold and Hadley had produced an over-sophisticated machine and that the men were giving an honest account. Whatever the truth of the matter it would certainly have been dangerous beyond words to have attempted to lower or raise the wheel unless it was equipped with a sluice gate which could be lowered to shut off the current from the wheel. Even so, unless there was also some sort of locking-device to hold the wheel in place, whatever its position, there was a clear risk that it could become in effect locomotive and begin to climb the rundle or spur-wheel or else pull on the capstans and revolve them like circular saws. But in fact Beighton's drawing omits so many features that the machine must have had, quite apart from those omitted for purposes of exposition, that any lengthy discussion of it would be quite otiose.

If we now recall the equation between Sorocold's lever and Scamozzi's *bilancia* the possibility is at least raised that Hadley was not the first in the field and that possibly one of his precursors was the anonymous engineer of the mill on the river Deman. Scamozzi talks of, and Ramelli shows, the whole body of a mill being moved up its legs by the force of screws, an extravagent waste of effort when the advantage it yielded could be secured by moving the waterwheel alone, a labour easily performed, as we gather from Beighton, if it were counterweighted.

Although the evidence for the use of sector and chain in this particular application is sketchy and late, there is rather more coherence in the case to be considered next. Ironically, it is one at some remove from the world of actual practice. Christiaan Huygen's *Projet de 1659 d'une Horloge à Pendule Conique*, and indeed the other projects having to do with chronometry with which he was occupied as late as 1693, reveal that sectors and chains, both with and without racks, deployed in a variety of ways, played an important role in his schemes.[38] I do not intend here to enter into the complexities presented by these materials except to observe that the presence of these elements in Huygen's work almost certainly indicates that devices or machines in common use were the sources from which they were derived. If one were to take, for instance, the sketch accompanying the *Projet* of 1659 (Figure 17) it is immediately apparent that its central feature is a sector device of exactly the form of MS.B. f54r.[39] It serves, however, merely as a balance. But did balances of this type then exist? The answer might well seem to be supplied by J. A. Schmidt in a small book published in Helmstedt about 1710 entitled *Theatrum naturae et Artis*. In a section devoted to weighing devices a balance like that used by

Inventum die 5 Oct. 1659

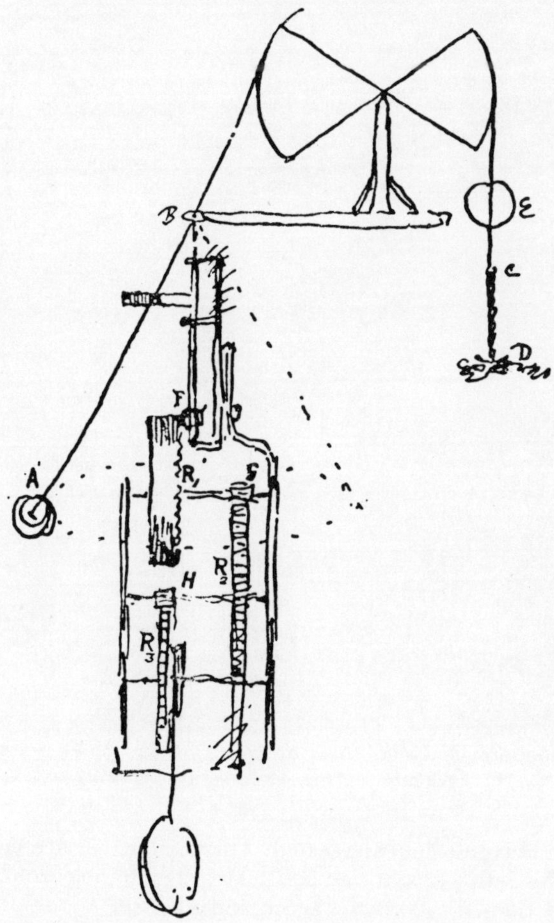

Figure 17. Suspension for conical pendulum, from C. Huygens, *Projet de 1659*, in *Oeuvres Complètes*, reprint Amsterdam 1967, Vol. 17, p. 88.

Huygens in 1659 is described as *libra communis* (Figure 18).[40] Huygens was to use the 'common scale' in 1675 and again in 1693 in his *balancier marin parfait*, features of which schemes reappear later in the work of chronometer makers such as Henry Sully and John Harrison. Huygens's correspondence also affords, as it happens, one other piece of evidence which suggests that sector and chain devices were, despite the paucity of direct evidence, in wide employment in the second half of the seventeenth century. Charles Perrault, writing to Huygens from Viry in October 1669, referred to a water-clock they had discussed which he evidently proposed to install in his grotto. He had decided, he says, to supply it with a

Figure 18. Common balance (*libra communis*) from J. A. Schmidt, *Theatrum Naturae et Artis*, Helmstedt, *c.* 1710.

pendulum as Huygens had suggested. The clock's balance beam which he sketched in his letter has at one end a weight hanging from a sector and chain.[41] Had the *libra communis* again supplied the model or should one look rather to some form of pendulum pump?

References to sectors, however, hard to come by before about 1700, become a veritable torrent in the eighteenth century and figure in an enormous number of mechanical ensembles. Altogether they offer a prospect of the eighteenth century's love of mechanical gadgetry which is both intrinsically interesting and of service as evidence that in respect of the pendulum pump and its derivatives Europeans were daily in Leonardo's debt whether as scientists using precision instruments, as artisans plying their trades, or as housewives preparing their country messes. Some of these uses will be looked at presently but as a preliminary it may be remarked that since they are so varied and wide-ranging in nature, it is hard to believe that they are all as genuinely late, from *c.* 1710 to *c.* 1775, in their adoption as on strictly chronological grounds they must appear to be.[42] Chance survival of evidence has no doubt played a large part in distorting the picture, but even so the nature of the evidence makes it possible to draw certain distinctions. It may well be the case that the employment of the sector and chain as an element in the construction of scientific instruments projected and actually built really is a late development, a function of the search for greater precision in measurement that characterized this field very noticeably from the later decades of the seventeenth century. It is, however, difficult to find any factor which will explain its late appearance in more prosaic situations unless it is that lack of direct evidence is distorting one's view of things.

In 1721 Henry Sully, a maker of chronometers, English by birth but who spent most of his life in France, began work on a marine time-keeper which he finally presented to the Académie des Sciences in 1724. As a controller Sully used a version of the sector and chain. A weighted lever, acting as a sort of horzontal pendulum, was equipped with a sector to whose lower end was attached the flexible cord which played between two curved cheeks. Two years later he published an account of it which contains an engraving of the arrangement with the inscription, 'Nouvelle Pendule à Levier approuvé par l'Académie Royale des Sciences 1724'.[43] In the same year as Sully's book appeared, Jacob Leupold published the third part of his *Theatrum Machinarum* in which he described various automatic recording machines which would trace continuous records, hour by hour, of variations in temperature and in barometric pressure, in both of which situations one finds balance beams with sectors and chains. There is also another machine in which a variant of the device acts like a *fusée*.[44] The use of the balance barometer with sectors and chains later spread to England where perhaps other instrument-makers than Watt had read Leupold's work. J. H. de Magellan, writing in 1779, claimed that he had himself made improvements in barometers of this type and that he had seen two such instruments, one by Adams, made, possibly in 1760, for George III, the other begun by Jonathan Sisson (d.1760) which he had himself improved.[45]

By the 1780s the balance beam, four feet long, with sectors and chains, appears in yet another role, as part of the gasometer constructed for Lavoisier by Mégnié following Meusnier's plans in 1783, and again in a later type constructed in 1787. The first of these gasometers had a chain, said wrongly by Daumas[46] to have been invented by Vaucanson, of a special type supposed not to be subject to elongation under tension. The shape of each individual link can best be imagined as being like that of a lyre with hooked ends (by which it hangs from the lyre above). Such a chain may well not have stretched but another desideratum is that the chain should lay itself flatly on the sector as the beam end rises and pay out smoothly from it as it descends. The pitch chain (or *chaine anglaise* as it was sometimes called in France, *Uhrkette* (clock chain) in German) might well seem a better type for use in precision situations.

Such are the uses of the device in laboratory work but no less diverse a picture is presented when one looks at the situation as far as the tasks of everyday life are concerned. I pass over its use, recorded in the *Encyclopédie*,[47] for the ringing of small bells, although no doubt, it could join that select collection of devices, the fly-ball governor, the vertical axle revolving bookcase, the hot air turbine and Villard de Honnecourt's turning angel, that have assisted in what Lynn White has called the technology of prayer. I wish instead to draw attention to another use, also illustrated in the *Encyclopédie*, that is in the preparation of lead sheet.[48] Attached to the preliminary forming table on to which the liquid lead is to be run is a trough holding about 3,500 pounds of metal. The forward lip of the trough is hinged to the top of the table so that when the back of the container is lifted by chains it spills its contents evenly over the casting bed. The success of the operation lay not only in establishing a level surface for the lead to flow over but also in precision of pouring, especially important since the metal, when poured, was not at its most fluid, having been allowed to cool to the point where a paper held over the metal would only turn brown. The sectors and chains of the counterweight beams ensured a uniform flow of lead over the lip of the trough since the pull would be immaculately vertical. Yet it would appear that the procedure, including the final rolling of the cast lead sheet, had been modelled on that in use in England since the beginning of the eighteenth century.[49] In fact a company had been formed in London to produce rolled lead as early as 1678. In England the sector and chain was later used about 1775 to ensure that candlewicks should be plunged with precision into tallow.[50] This ensured an equal build up of material around the wick and consequently a candle that would burn uniformly and without waste. From the end of the chain hung a horizontal rod with a row of hooks over which were placed the loops of the cotton wicks. The wicks would already have been well primed with tallow so that one must imagine them hanging down straight like rods (Figure 19). The operator standing next to the trough full of tallow, which like the lead would be on the point of losing its fluidity, pulled on the beams which, swinging down, allowed the wicks to receive their coating of tallow. After a brief immersion the operator released the beam which swung smoothly back, lifting the wicks clear of the trough where they would drip off and harden in the air and thus be ready for their next coating. Nothing like this is to be seen in Duhamel de Monceau's 'L'Art du Chandelier'[51] for there everything is done by hand, but the superior speed and ease of the English method of production would not have been gained at the expense of quality; rather the reverse.

The English iron industry also was to exploit the sector and chain. Isaac Wilkinson's third patent, No. 713 of 1757, for 'a machine or bellows to be wrought by water or fire-engines' proposed its use in a cylinder blowing machine for blast furnaces in place of the cuneate form of bellows then almost universal. The arrangement overall is reminiscent of the Newcomen engine, except that air was to be pumped instead of water.

Figure 19. Candle dipper, *c.* 1800. By courtesy of the Castle Museum, York.

Isaac hoped to retrieve his own fortunes with the idea and may well have succeeded, for a cylinder blowing-engine was in use at Walker's ironworks in Rotherham by 1762 and was the type the Carron directors insisted on (in the face of Smeaton's desire to design his own) when the time came to build blast furnaces Nos. 3 and 4.[52] William Wilkinson installed steam driven cylinder blowers in 1785 at the Mont Cenis ironworks in Burgundy. Johann Ferber, a good judge, was staggered at their enormous force when he visited the works in 1788.

Perhaps one would hardly expect the sector and chain device, pervasive though it has now been shown to be, to have invaded even the kitchen parlours of the eighteenth century, but such was the case. The open fire of the domestic hearth was, of course, where food was prepared and the wrought-iron accoutrements that were so necessary a part of its furniture have a long and fascinating history. The andirons, cats and spits deserve description, but I wish to draw attention to the chimney-cranes which were a universal feature of English, and no doubt European, hearths at this time (Figure 20). A housewife tending her pot needed always to be able to adjust its position in three ways. The first movement the crane had to be capable of was to turn on its vertical axle and thus bring the horizontal arm and everything hanging from it clear of the fire so that the housewife could perform her offices. Next she had to be able to adjust the rate of cooking by varying the height of the pot above the fire, and finally, when all was ready and the stew needed only to be kept nicely warm until dinner time, a horizontal movement was necessary that would bring the pot to the side of the hearth. Any blacksmith in eighteenth century England would have quickly made a crane which would perform the first movement, but generally speaking the others would be left to the cook to manage with her bare hands, with a certainty of soot and a good chance of being burnt as well. The sophisticated three-movement crane took these horrors out of cooking. A balance beam mounted on a travelling carriage permitted motions two and three to be given to the pot from a safe distance.[53]

It remains to say something about the Newcomen engine of 1712 which perhaps more than any other of the machines that have been considered would appear to have been influenced by its Leonardian antecedents. In 1963, in honour of the tercentenary of Thomas Newcomen's birth, Dr. Joseph Needham presented a paper to the Newcomen Society called 'The Pre-Natal History of the Steam Engine'[54] and revealed the complexities of the currents of design that lay behind, to use Usher's words, 'the greatest single act of synthesis' in the development of steam power. In many ways the metaphorical colouring of Needham's title is appropriate because a succession of commentators ever since 1725, when the author of the poem 'The Prize Enigma' presented his version of the engine's ancestry, have described the engine in a curiously anthropomorphic and anatomical fashion:

> I now can raise my hand above my head;
> And now, at last, I by myself am fed.
> On mighty arms alternately I bear
> Prodigious weights of water and of air.[55]

But Needham's anatomizing neglected to mention, if it is permissible to continue the trope, the arms and shoulder blades of the machine, for he had nothing to say of the sectors and chains nor indeed of the swape beam itself. In another sense too his analysis was, I think, incomplete, if not

Figure 20. Chimney crane, *c*. 1800. By courtesy of the Castle Museum, York.

actually misleading. All the elements are presented, as it were, in a heap much as a cook might assemble all her ingredients, but the situation Newcomen faced about 1698, or whenever it was precisely that he began work on his machine, was obviously far different. Beyond doubt all of the many ingredients were already gathered together in other syntheses such as this survey has already considered. If, as appears overwhelmingly likely to have been the case, Newcomen knew at least of the working principle of Papin's engine of 1690, and the review of Papin's *Recueil* which appeared in Volume XIX (1697) of the *Philosophical Transactions* would have acquainted him with its essential features, he would have sought for a machine which could be adapted in such a way as to permit the weight of the atmosphere to be exploited. He had no need to conjure up a whole machine for this purpose *ex nihilo*. Consider the Leonardian machine of MS.B. f54r (Figure 1) which was the starting point of this enquiry. Is it really possible to believe that Newcomen had not seen English versions of it at work in Devon? This is almost a rhetorical question, for English colonists in North America had already taken the machine with them across the Atlantic. The issue for June 1755 of the *Gentleman's Magazine*[56] contained an article acquainting the English public with the equipment needed and the procedures followed by the planters of the southern colonies in processing their crops of indigo. The engraving reproduced by G. Terry Sharrer in his article 'The Indigo Bonanza in South Carolina' shows that the installation was all very simple, nothing more than a pump and some vats for steeping the plants in.[57] But what a pump, and of all conceivable pumps what an astonishing one to find in use in South Carolina in 1755. Two and a half centuries after Leonardo had first sketched his pendulum pump its exact replica makes its appearance in an English magazine. It is, of course, possible that Newcomen, searching for a machine on to which he might graft the new idea, sought other models. Perhaps he had seen the sort of pump that Desargues and La Hire (and Mandey) had made familiar, and I suppose it is not impossible that if Watt had had his kettle, Newcomen could be supposed to have had a three-movement chimney crane. To mention such possibilities is to see at once that they lead nowhere. It might seem rather that the marriage of Papin's cylinder and Leonardo's pendulum pump lies at the heart of this engine, and everything else, snifting valve, cataract and so on, the technical problems included, takes its place within the context of his first basic synthesis. Once the possibility of obtaining power from an active piston was available, the pendulum would naturally disappear. The one-stroke nature of the device would necessitate the loading of the pumping end of the machine and thus far one would guess that Newcomen found his progress gratifyingly rapid. Thereafter, if Tricwald's account of the accidental discovery of internal water injection is correct, a long period followed while Newcomen cast about for a method by means of which a speeding up of the cycle might be induced, and thus finally bring Papin's idea to a practical delivery. Here, however, the world of technics had nothing much to offer either immediately or via any conceivable process of *gradual* evolution.

Now that these remarks on Newcomen's engine have permitted a resumption of the development of pumping engines, one further example ought to be considered before it is brought to a close. In 1711 Mathias Höll began construction of six *Stangenkünste* at Windschacht, all of which embodied the sector and chain (Figure 21). The context of his work has already been considered elsewhere,[58] and here it will suffice to say that his complex arrangement was very probably modelled on a simpler system used in rod-engines working single lines of field-rods.

What appears to be a sudden efflorescence in the use of the sector and chain device as the eighteenth-century advances may be no more than an accident of survival of evidence, while if it is not, it might be held to reveal something perhaps, of the accelerating pace in the development of engineering skills which would accord well with the orthodox view of the relatively rapid onset of industrialization in the second half of the century. Although it would be inappropriate to discuss such larger issues here, it will not do to suppose that the sixteenth and seventeenth centuries saw little enlargement of the inventories of engineers. Perhaps the question is

Figure 21. Mathias Höll's Rod-machine of 1711 (*Stangenkunst*) with sectors and chains, from N. Poda, *Kurzegefasste Beschreibung*, Prague 1771.

less whether the evidence for eighteenth-century development has survived than whether that for the previous two centuries has been to a very large extent lost. A fairly recent survey of the historiography of the Industrial Revolution rejects the idea of a starting point or discontinuity some time after 1750 and prefers to think of that period instead as 'the culmination of a most unspectacular process, the consequence of a long period of economic growth'.[59] If this non-heroic view of the matter is somewhere near the truth then it seems not unreasonable to suppose that a parallel growth, either as cause or effect, was taking place in the field of mechanical engineering. The machine books, however unequal in value and however plagiarizing they might sometimes be, indicate something of this development, and the growing realization of what machines had permitted to be done and might yet permit: a feeling of *vivitur ingenio*. Machine histories, as far as they can be recovered, provide the concrete evidence which shows such optimism to have been well founded.

Postscript

In discussing the moment when Newcomen began work I suggested that Leonardo's pendulum pump (Figure 1) might well have played an important part in providing him with a model. Since writing this the award of a British Academy exchange scholarship has made it possible for me to visit (in May 1978) the central mine archive at Banská Štiavnica in Slovakia. I was fortunate enough to discover there a document, of the late seventeenth century, which suggests the possibility that Newcomen may have had a rather more sophisticated model on which to base his experiments. The material in question (Inventory No. 10,641) consists of a four page report of a water-driven mine-ventilating machine already in use in the Heuffenthal valley in the Harz mountains. In one version of this machine two beams with sectors and pitch chains (very like two Newcomen engine beams) are set in motion by a water-wheel with two cranks. The kinematics of the machine are precisely those of the Newcomen engine: the down throw of the crank (corresponding to the power stroke of the steam engine) pulls down the beam whose return is effected by the counterweight of the pump rods. Although machines similar to this were in use in Scottish mines in the 1730s I had always supposed them hitherto to have been hydraulic imitations of the Newcomen engine. It might well seem now that the reverse is nearer the truth of the matter.

Notes

1. The *locus classicus* is, of course, a footnote in Karl Marx's *Capital* where a plea for a history of the productive organs of man in society is prefaced by reference to Darwin's work on what Marx called 'the history of natural technology'. *Capital* (Everyman ed.), Vol. 1, London 1957, p. 392, note 2.

2. R.U. Sayce, *Primitive Arts and Crafts: an introduction to the study of material culture*, Cambridge 1933 (rev. ed. New York 1963). His object, as he says in the preface to the second edition, 'was to introduce . . . some of the principles that can be seen to operate in the invention, diffusion and general evolution of human artifacts'.

3. Sayce's remarks (*op. cit.*) on this subject are very revealing especially those concerned with the phenomenon of retention of redundant features in an artifact because they seem to the maker to constitute part of its essential being, so powerful is the hold of the prototype. He gives examples of skeuomorphs from various cultures. L. Reti, 'Francesco di Giorgio Martini's Treatise on Engineering and Its Plagiarists', *Technology and Culture*, Vol. 4, 1963, pp. 287–98, has drawn attention to a similar sort of situation in art, and one long familiar to art historians, and has suggested that certain 'images' in technology may well have had a similar compelling force on the imaginations of engineers.

4. C. Maltese and L.M. Grassi (eds,), *Trattati di Architettura, Ingegneria e Arte Militare*, Vol. 1, Milan 1967, f45v. Martini's pierced piston rods fitted with anti-friction rollers to accommodate the circular motion of the swape beam did permit force to be exerted on both strokes of the piston. However, a more exact comparison would be with Leonardo's design MS.B. f20r. As far as I know Francesco's device, if it was his, did not survive beyond the sixteenth century.

5. Watt's parallel motion device was first used at the Albion Mill, Blackfriars Bridge, London, when the mill began working in March 1786. No drawing exists as far as I know of Wasborough's first engine erected in Bristol in 1779. Since it was his second, built for Pickard at Snowhill, Birmingham, that had a rack removed in 1780 in favour of a crank and connecting rod (on the non-engine end) and since Watt only solved the question of positive linkage for the engine end some time later, it is conceivable that Wasborough's first engine may have borne as strong a resemblance to MS.B. f20r as Newcomen's does to MS.B. f54r.

6. An almost certain example dates from 1582 but the device was definitely in use well before 1652.

7. Mariano di Jacopo detto Il Taccola, *Liber Tertius de Ingeneis ac Edifitiis non usitatis*, (ed. J.H. Beck), Milan 1969, ff 40, 41. Taccola's devices were certainly common property by Leonardo's time among Italian engineers. Taccola's special form of scaling ladder working in the same way as the manganon appears in Leonardo in modified form in *Codex Forster*, II, f46v.

8. This becomes evident if one sketches the machine in its positions before and after firing. Two quarters are to have rope paid out from them, one quarter has the rope secured to it.

9. C. Truesdell, *Essays in the History of Mechanics*, Berlin, Heidelberg, New York, 1968, p. 20, 'Like none else could he see', and p. 62, 'He shows us motions'.

10. L. Reti, 'The Leonardo da Vinci Codices in the Biblioteca Nacional of Madrid', *Technology and Culture*, Vol. 8, 1967, pp. 437–45. Leonardo's experiences with machine design would seem to demand a sober practical assessment of their value set against an analysis of existing procedures.

11. L. Reti, 'The Problem of Prime Movers', in *Leonardo's Legacy*, ed. C. O'Malley, Berkeley 1969. But see more particularly pp. 83–6. I must differ, however, from Reti's interpretation of MS.L. f76v as my reading of the excavator's action will indicate.

12. *Ibid.*, p. 85.

13. L. Reti, 'Leonardo and Ramelli', *Technology and Culture*, Vol. 13, 1972, p. 605.

14. Morris's name is variously spelt: Morice (Stow), Moris (Patent).

15. Rhys Jenkins, *Collected Papers*, Cambridge 1936, p. 140.

16. Fynes Moryson, *Itineraries*, London 1617, p. 11. In Part III, p. 91, Moryson, who was of course a Dutch (or Deutsch) man, explains that the name Dutchman was applied indifferently to both Germans and Hollanders because their language and manners were so similar.

17. J. de Strada, *Künstlich Abriss*, Vol. 1, Frankfurt, 1617, fig. 43. The caption runs: 'Ein machina durch hülff eines Menschen und eines Gewichts und zwei Pompen das Wasser in die höhe zuheben'. The advantage of Strada's form of the machine is presumably constructional (I refer to the uncut top of the wheel). If, as in Morris' upper wheel, the top were not cut away, a band of iron could be placed round the upper part of the curve and thus provide a good anchorage for the end links of both the pump chains. G.A. Böckler, *Theatrum Machinarum*, Cologne 1662, takes his figure 102 directly from Strada.

18. P. Skippon, 'An Account of a Journey made thro' part of the Low Countries, Germany, Italy and France', p. 464, in J. Churchill, *A Collection of Voyages and Travels*, Vol. VI, London 1732. Skippon was in Germany in June 1663.

19. A feature that makes one suppose Morris' model may have resembled Strada's pump. As for the double or reverse chain, it remained long in use. Barney's engraving, published in 1719, of the Dudley Castle engine (Newcomen's first engine of 1712) shows the plug rod given a positive downward drive in this way. John Smeaton used it for his Borough Wheel (London Bridge) pumping engine of 1767.

20. W. Schildknecht, *Harmonia in Fortaliciis, Construendis, Defendendis et Oppugnandis. Das ist: Beschreibung Festungen zu Cawen*, Part III, Stettin 1652, p. 116, Plate Z, No. 1.

21. *Ibid.*, Part III, p. 117.

22. A. Ramelli, *Le Artificiose et Diverse Machine*, Paris 1588. There are other variations on this theme but No. 32 may stand for the type.

23. W. Schildknecht, *op. cit.* Three pumps are shown in plate Z, p. 117, one on plate A, p. 120. Little appears to be known of Schildknecht's biography. The title page of his book speaks of him as one time engineer to the Prince (Fürst) of Pomerania and as still engineer, ordnance master and surveyor to the city of Stettin where his book was published. J.H. Zedler, *Grösses Universal Lexicon*, Vol. 34, Halle and Leipzig 1742, p. 1545, has little to add to this except to misquote the title of his book. However, Zedler's separate entry for Wendelin's son Christian (1626–79), successively major and colonel in the private guard of the Duke (Herzog) Gustav Adolphe of Mecklenburg-Gustrow, helps perhaps to establish some idea of Schildknecht senior's floruit.

24. Sir Edward Ford's machine at Somerset House, London, sketched by B. de Monconys in 1663 and published in his *Journal de Voyages*, Vol. 2, Lyon 1666, p. 29, fig. 3, hardly solved the problem of how best to do this but shows what an amateur was likely to come up with. The evolution and history of the face-wheel has not so far as I know been attempted but what would be elements in such a history spring readily to mind: Besson's water-driven face-cam working a swape, of 1579, reappears turned upside down and driven, not driving, in James Watt's first proposal of 1781 for converting the reciprocating action of the steam engine into rotary motion. Obviously the idea remained in circulation between these two dates to offer Ford a hint.

25. R. Taton, *L'Oeuvre Mathématique de Girard Desargues*, Paris 1951, p. 64. Although

nothing is known of what he communicated, it was Desargue's lifelong habit to give free instruction to tradesmen and craftsmen with whom he would also share his ideas. Desargues believed that by contributing to technical progress he would do most to ameliorate the lot of the poor.

26. Christiaan Huygens, *Oeuvres Complètes*, Vol. 7, Correspondence (1670–5), The Hague 1897, Letter No. 1844, 22 Sept. 1671.

27. *Ibid*, Letter No. 1850, 29 Oct. 1671.

28. His presence in Paris in 1658 has been established from a will he made in that year, cf. J. Balteau, *Dictionnaire de Biographie Française*, Vol. 10, Paris 1962, pp. 1183–4.

29. His *Traité de Mécanique*, Paris 1695, Proposition CXV, pp. 368–74, contains the same description.

30. B.F. de Belidor, *Architecture Hydraulique*, Vol. 2, Paris 1739, Part 1, Ch. 4, p. 161, mentions that the swape beams are 30 feet long, the rollers 8 inches and the waved wheel 7 feet in diameter. Such a pump would lift water 150 feet at the rate of over 1,250 gallons an hour (my rendering of Belidor's 5,500 pintes or $19\frac{1}{2}$ muids per hour).

31. *Traité des Epicycloîdes*, Paris 1694, Proposition IX, pp. 68–72: 'Construction d'une machine pour élever de l'eau sur la forme précédente'. A necessary work, Le Hire explains, because 'je n'ay point sçu que cet excellent Géomètre eut jamais rien expliqué de sa construction . . . je crois qu'il en avoir seulement determiné la figure méchaniquement'.

32. Book 10: of epicycloids and their use in mechanics.

33. *Ibid*., p. 297.

34. *Op. cit.*, plate 7, figs. 1, 5, 6 and 8.

35. V. Scamozzi, *L'Idea della Architettura Universale*, Vol. 2, Venice 1615, p. 370. L. Reti, in 'The Problem of Prime Movers', traced the diffusion of the idea from Leonardo. Juanelo Turriano appears to have used such a wheel at Toledo in 1569; at least, the words used of the wheel there, that 'la creciente del rio no puede impedirlo', seem to suggest this. See L. Reti, *El Artificio de Juanelo en Toledo*, Toledo 1967, p. 19, quoting Zuccaro. The use of screws to lift wheels continued after Scamozzi's time but in the two examples known to me they were provided only in case the wheel got damaged and needed to be lifted clear of the water for repair. Belidor, *op. cit.*, Vol. 1, Part 1, Ch. 1, plate 3, shows an engraving of such a mill at Mont-Royal on the Moselle. His description will be found on p. 296. The wheel of Laurent's pumping machine at Pontpean, Brittany, built in 1755, could also be lifted by screws in case of damage. A full description is given in the *Encyclopédie*, Vol. 13, Paris 1765, p. 10, under 'Pompes'. Perhaps Leonardo had intended nothing more in the first place.

36. Celia Fiennes, *Journeys*, London 1888, p. 140 (London 1948, p. 170). The best short account of George Sorocold's career is by F. Williamson in *Derbyshire Archaeological and Natural History Society*, N.S. Vol. X, 1936. Sorocold died of his injuries c. 1717 after falling into the wheel-race at Cotchet's Silk Mill, Derby.

37. *Philosophical Transactions*, Vol. 37, 1731, No. 417, pp. 5–12 and plate 1.

38. C. Huygens, *Oeuvres Complètes*, reprint Amsterdam 1967, Vol. 17, pp. 88ff. (scheme of 1659) and Vol. 18, pp. 501ff. (schemes of 1665 to 1693).

39. *Ibid*., Vol. 17, p. 88, fig. 13.

40. The title pages of both editions of the book are undated. The author, signing himself I.A.S.D., has been identified as Johann A. Schmidt. The British Museum catalogue tentatively gives the date of publication of both editions as 1710. It might here be noticed also that John O'Kelly in his letter of 1725 was to speak of the beam and sectors of the Newcomen engine as being exactly like an ordinary balance.

41. C. Huygens, *Oeuvres Complètes*, Vol. 6, Correspondence (1666–9), The Hague 1896, Letter No. 1769, 28 Oct. 1669.

42. Not discussed here are double ropes as an aid to the steering of men-of-war which came into use almost simultaneously in the English and French navies soon after 1700; double chains employed by Richard Newsham in his fire-fighting pumps of 1721 and 1725, the single sector and chain used by George Gerves in his pumping machine of c. 1730, and by William Henry in his sentinal register stove of 1771. After 1771 references become even more abundant.

43. The plate is reproduced in R.T. Gould, *The Marine Chronometer*, London 1960, pp. 35–6.

44. Jacob Leupold's meteorological recording devices of 1726 exemplify very well a rapidly developing expertise: *Theatrum Statici Universalis, sive Theatrum Aero-Staticum . . .,* Leipzig 1726, plate XXIII, figs. II, III and IV. For a description of Leupold's machines and reproductions of these engravings see H.E. Hoff and L.A. Geddes, 'The Beginnings of Graphic Recording', *Isis*, Vol. 53, 1962, pp. 287–324.

45. Magellan's balance barometer is reproduced in W.E.K. Middleton, *History of the Barometer*, Baltimore 1964, p. 103, fig. 6.9.

46. M. Daumas, *Lavoisier: Théoreticien et Expérimentateur*, Paris 1955, Ch. VI, p. 146. His fig. 5, plate III, does not show the chain clearly but it is distinctly shown in J.B.M. Meusnier, *Description d'un Appareil propre manoeuvrier differentes Espèces d'airs*, Vol. 2, Strasbourg 1787, p. 432. J. Leupold, *Theatrum Machinarum Hydraulicarum*, Vol. 1, Leipzig 1724, p. 52, shows three types of chains including the lyre-shaped variety. Vaucanson (b.1709) was then fifteen years old. But in any case they are to be seen in G. Agricola, *De Re Metallica*, Basel 1556, Bk. VI.

47. *Recueil de Planches*, Vol. 5, Paris 1767, plate VIII, fig. 9, of Fonte des Cloches.

48. *Ibid.*, Vol. VIII, Laminage de Plomb, plates 2 and 4. For a description of the process see *Encyclopédie*, Vol. IX, Paris 1765, p. 230.

49. P. Rémond de Saint-Albine, *Mémoire sur le Laminage de Plomb*, Paris 1731, p. 45 (Query No. 1).

50. The Castle Museum, York, has preserved an entire tallow candle maker's shop. The candle dipping machine is so beautifully counterweighted that an exquisite degree of control is possible. More importantly its use eased the operative's work while permitting production to be increased by over sixty per cent. A. Rees, *The Cyclopaedia*, Vol. VI, London 1819, sub *Candles, making of dipped*, states, somewhat vaguely, that such machines had been introduced 'within 15 or 20 years past'. The section in question is, however, likely to have been written as early as 1795. For Ree's machine see *Plates*, Vol. II, 1820, *Candle Making*, figs. I and II.

51. Duhamel de Monceau, 'L'Art du Chandelier', *Description des Arts et Metiers*, Vol. 1, Paris 1764. Producing *chandelles plongées* was an extremely unpleasant job, especially the last part which involved forcing their bases against a hot plate so as to ensure that they would stand upright.

52. H.W. Dickinson reproduces a drawing of the machine in *John Wilkinson, Ironmaster*, Ulverston 1914. The layout, reduced to two cranks, and wrongly ascribed to John Wilkinson, is diagrammatically represented in J. Needham, *Science and Civilization of China*, Vol. 4, Cambridge 1965, Part 2, p. 380.

53. The Ironwork Gallery in the Victoria and Albert Museum, London, has one extremely elegant example and the remains of another. The example in the Castle Museum, York, is decidedly a heavy duty type, as will be evident from Figure 20. L.A. Shuffrey, *The English Fireplace and its Accessories*, London 1912, p. 62, fig. 69, shows the type reduced to essentials. See my article 'Sophisticated Cranes', *The Connoisseur*, 1974 (June) for further details.

54. *Transactions of the Newcomen Society*, Vol. XXXV, 1962–3, pp. 3–58, reprinted in *Clerks and Craftsmen in China and the West*, Cambridge 1970, pp. 136–202.

55. Originally printed in *The Ladies Diary*, London 1725, but quoted in L.T.C. Rolt, *The Pre-History of the Steam Engine*, London 1963, p. 6.

56. Vol. 25, p.259. The caption to the engraving merely records 'Two pumps in a frame worked by a pendulum'. A carpenter would be needed, the writer adds. Altogether it would seem that note 1, pp. 172–3 of Lynn White's *Mediaeval Technology and Social Change*, Oxford 1963, stands in need to revision.

57. See *Technology and Culture*, Vol. 12, 1971, p. 451, fig. 1.

58. See my article 'The Vocabulary of Technology', in *History of Technology: Second Annual Volume, 1977*, eds. A. Rupert Hall and Norman Smith, London 1977, pp. 125–55.

59. R.M. Hartwell, 'The Causes of the Industrial Revolution: an essay in methodology', *Economic History Review*, 2nd Series, Vol. 18, 1965, p. 180. An *obiter dictum* of Alfred Marshall quoted by Hartwell is equally apposite, 'economic evolution is gradual and continuous in each of its numberless routes'.

The Contributors

P.S. BARDELL is Senior Lecturer in mechanical engineering at the Colchester Institute of Higher Education, Essex.

K.R. FAIRCLOUGH is a former extramural student of the University of London at Walthamstow Adult Education Centre. He has written on the Industrial Archaeology of the Lea Valley and its documentary history.

ROBERT FRIEDEL is a research worker at the National Museum of History and Technology, Smithsonian Institution, Washington, D.C. U.S.A.

J.G. JAMES is a Principal Scientific Officer at the Transport and Road Research Laboratory, Crowthorne, Berkshire. He has a special interest in the history of structural ironwork.

L.J. JONES is Senior Lecturer in mechanical engineering at the University of Melbourne, Victoria, Australia and was an Academic Visitor at Imperial College, London during 1978/9.

G. HOLLISTER-SHORT is Principal Lecturer in History at Shoreditch College, University of London Institute of Education and Honorary Lecturer at the Department of History of Science and Technology, Imperial College, London.